T0145979

Lignites: Their Occurrence, Production and Utilisation

Lignites: Their Occurrence, Production and Utilisation

Clifford Jones

Federation University Australia

Whittles Publishing

Published by
Whittles Publishing,
Dunbeath,
Caithness KW6 6EG,
Scotland, UK
www.whittlespublishing.com

ISBN 978-184995-180-7

Print managed by Jellyfish Solutions

Dedicated to the memory of my father

Jack Bryan Jones

for the centenary of his birth

Contents

CHAPTER 4

Electricity generation I – Germany

CHAPTER 5

Electricity generation II – Other European countries

CHAPTER 6

Electricity generation III – North America

CHAPTER 7

Electricity generation IV – Asia

CHAPTER 8

Electricity generation V – The Indian sub-continent

CHAPTER 9

Electricity generation VI – The Former Soviet Union

CHAPTER 10

Electricity generation VII – Australia

CHAPTER 11

Briquettes

CHAPTER 12

Carbonised products

CHAPTER 13

Gasification

CHAPTER 14

Conversion to liquid fuels

CHAPTER 15

Chemical substances from lignites

CHAPTER 16

Unworked lignite deposits

CHAPTER 17

Hazards with lignites

CHAPTER 18

Leonardite

CHAPTER 19

Examples of carbon capture and storage (CCS) at lignite-utilising plants

CHAPTER 20
Co-combustion of lignites with other fuels

CHAPTER 21
Comparisons with peat

CHAPTER 22
Comparison with sub-bituminous coals

Preface

The writing of this book has been a deeply satisfying experience. Lignites are a fuel resource upon which there has been heavy reliance for a long time in several parts of the world, and a comprehensive text on them at a time when there is such urgency on two fronts – awareness of the finiteness of energy reserves and the need to control greenhouse gases – is timely and, indeed, a welcome challenge to the writer.

The research literature has been extensively drawn on, and the fact that the literature contains a good deal of recent work on lignites is evidence of the place they have in world energy supply in the second decade of the 21st century.

Thanks are due to Whittles Publishing, with whom I have by now had a very good working relationship for 20 years. I also acknowledge advice from Professor Roy Jackson and Dr Marc Marshall.

Clifford Jones
Churchill, Victoria

CHAPTER 1
THE NATURE OF LIGNITES

1.1 Lignite in popular conception

Lignite is also known as brown coal, and a comment on the equivalence or otherwise of the two terms is made later in the chapter. Irrespective of formal definitions it is a fact that lignites are brown, and it sometimes surprises people to be informed not only that there is coal which is brown in colour but that there is heavy dependence on it in many countries.[1]

There are differences other than in colour between brown coals and black coals and an account of these follows. It will be preceded by the important information that lignite accounts for 23% (tonnage basis) of the known coal reserves of the world [1].

1.2 Background on coal formation

Coalification is the process by which, on a timescale of tens or hundreds of millions of years [2], vegetation is converted to coal. The maturing of coal on this timescale from one rank to another is termed the coalification sequence, details of which are well known. A graphic representation of it can be found in [3] and summarised: vegetation → peat → lignite → sub-bituminous coal → bituminous coal → anthracite.

The first step is that vegetation becomes immersed in water where, by the action of microorganisms, it is converted to peat. Biochemical action ends there and factors in the advance along the coalification sequence are pressure, temperature and time. The next stage beyond that is lignite a.k.a. brown coal and the subject of this book. The coalification sequence from vegetation to anthracite indicates how compression accompanies increase in coal rank. That is the meaning of 'rank' in coal science – degree of advance along the peat-to-anthracite sequence. Hence a lignite is a low-rank coal and a bituminous coal a high-rank one. In some conventions the terms are used in their comparative forms: a lignite is a low*er* rank coal and a bituminous or anthracite a high*er* rank coal. The term 'low-rank coals' used generically takes in sub-bituminous. In a few countries, as will be mentioned in the text, a distinction is made between brown coal and lignite but this is the exception rather than the rule, and a reader who wishes to do so will be able to confirm that the two terms are widely used synonymously.

When a lay person is informed that there is such a thing as brown coal, to enquire whether it is 'like peat' is a natural and indeed an intelligent response. The difference is clear enough on close comparison of the respective substances. The plant material evident in the peat is present in lignite but with a changed appearance, having undergone what some coverages of the subject call 'organic metamorphism' [4], itself an effect of the compression referred to. This takes the discussion into the realm of coal petrog-

1 Black coal is a recognised term in coal science, denoting coals from bituminous upwards in the coalification sequence, that is, bituminous to anthracite.

raphy which, as it relates to lignites, has a chapter to itself later in the book. There is return to this theme in section 21.2.

1.3 Brown coals vis-à-vis black coals

1.3.1 Calorific values

Differences are major and include the highly important property of calorific value. Selected values from the literature are given in Table 1.1 and must be viewed in the light of the following. A brown coal in the 'bed-moist' state, that is, in the coalfield before mining, will be at least 40% moisture and possibly as high as 65%. Water loss simply by natural evaporation on standing after being mined will double the calorific value, or more than double it. With any rank of coal, the calorific value depends on the ash content.

Table 1.1 Calorific values of selected coals.

COAL	CALORIFIC VALUE
Victorian brown coal, bed-moist [5]	< 10 MJ kg^{-1}. 65% moisture in the raw state (see comments on moisture in text)
Lignite from Coal Creek, North Dakota, USA [6]	Coal received at 38% moisture dried to 29% moisture. Calorific value raised from 6200 to 7100 Btu lb^{-1} ≡ 14.5 to 16.6 MJ kg^{-1}
Lignites from Mae Moh, Thailand [7]	In the range 9.1–12.7 MJ kg^{-1}. Ash contents in the range 13–45%
Sub-bituminous coal from Powder River Basin, USA [8]	8683 Btu lb^{-1} ≡ 20 MJ kg^{-1}
Pittsburgh bituminous [9]	28.5 MJ kg^{-1}. Value corresponding to a sample 9% in ash
German anthracites [11]	27–35 MJ kg^{-1}
Iranian anthracite [12]	32.6 MJ kg^{-1}
Tuncbilek lignite, Turkey [13]	20 MJ kg^{-1}

Victorian brown coals have two features which to some degree offset the high moisture in potential for usage and marketability: they are low in ash and in sulphur. The contents of the second row of the table require no comment beyond that non-SI units are still prevalent in the USA. Moving on to the next row, the Thai lignites were only up to about 15% water as tested. Clearly the low calorific value is due to the extremely high ash. There is more on the lignites from Mae Moh in Chapter 7. The Powder River Basin is partly in Wyoming and crosses the state border with Montana. The coal from there is 6.6% in ash and this is one of the factors in the quite high calorific value. Higher still is the value for a Pittsburgh bituminous coal. Up to and including the penultimate row the entries in the table are in ascending order of rank, and the increase in calorific value with rank is clear although this trend can be obscured by uncommonly high ash. There are bituminous coals of 15% or more ash content, notably in South Africa [10]. The table continues with the range of values for anthracites occurring in Germany. That for the anthracite in the following row is as high as one will encounter. The moisture content of the Turkish lignite which features in the final row had been reduced to 10% from the bed-moist state.

It is evident that ash is a major factor in the calorific value of coal of any rank. It derives from minerals and inorganics in the coal. The plant life from which coal is derived contains a small proportion of inorganic material, and to that is added minerals present in the environment in which decomposition of the vegetation took place [14]. Anticipating the next chapter, uptake of minerals occurs during 'maceration' (see section 2.1 for a definition).

1.3.2 Hardness

A contrast between brown coals and black coals almost as marked as the obvious one of colour is that the former are soft and the latter hard: 'soft coal' and 'hard coal' are quite acceptable terms distinguishing lignites from bituminous coals or anthracites. The hardness of any particular coal is determined quantitatively as the 'grindability', measurable as the Hardgrove index [15]. 'Grindability' is perhaps the more useful term, as many coals are destined to be burnt as pulverised fuel with particle sizes of ≈ 100 μm.

In the conceptually and experimentally simple Hardgrove index determination a known weight of the coal under examination is subjected to a specified amount of energy by manual application of a ball mill. Size analysis of the resulting particles is carried out, and the coal having been tested is placed on a scale from 30 (high resistance to grinding) to 100 (low resistance to grinding). The standards bodies (see Chapter 24) make available for purchase materials of particular Hardgrove index for calibration [16]. Table 1.2 gives some Hardgrove indices from lignites and some from hard coals for comparison.

Table 1.2 Hardgrove indices.

COAL	HARDGROVE INDEX
TX lignite [17]	48
Victorian brown coal briquettes [18][2]	40–45
Anthracites, in general [19]	35–40
Indian lignite 'as received' [20]	84
Kwa Zulu Natal, South Africa, anthracite [21]	57
Set of 30 Turkish lignites [22]	30–66

The value given for 'TX lignite' is one which has been taken to apply approximately to all lignites in Texas, and in [17] it is pointed out that for any coal the Hardgrove index has a significant dependence on moisture content. The remarkably high value for the Indian lignite should be seen as reflecting its 'as received' condition, which means that

2 Briquetting is compaction of crushed coal and moulding to make coal pieces of regular shape, and is fully discussed in a later part of this book.

large quantities of moisture had been retained. This figure, of intrinsic interest though it is, cannot be compared with the figure for Victorian brown coal briquettes where there would have been moisture loss before (by evaporation) and during (by pressure) the briquetting process. The fifth entry in the table is an example of application of one of the certified standards referred to above and the value is unexpectedly high for an anthracite. It was chosen for inclusion as an example of such an anomaly. Note that the range for 30 Turkish lignites in the following row goes beyond the value for the South African anthracite in the previous row. Yet there is no consensus amongst coal scientists that the Hardgrove index is imprecise. Quite the contrary! Several standards bodies have made it the basis of standards. The negative correlation of the index with coal rank is not as evident as intuitively expected and, as recognised by the standards bodies in the sample preparation which they require, a great deal depends on the condition of a coal when tested. Hardgrove indices of lignites feature later in the book, for example for Serbian lignites in section 5.8.4. The Hardgrove index as it relates to the performance of a coal mill, for the preparation of pulverised fuel (p.f.) from pieces of lignite, features in section 4.2.8.

1.3.3 Thermal decomposition

Moving left to right along vegetation → peat → lignite → sub-bituminous coal → bituminous coal → anthracite, volatile matter yield on heating diminishes. Vegetation will lose most of its weight on heating to say 800°C under conditions precluding burning (of which more later): anthracite will lose something like 5% of its weight in such a heating trial. This section is concerned with brown coals compared to black coals in this regard. The material lost on thermal decomposition is called volatile matter or simply volatiles. Moisture in the coal is not classified as a volatile (it is sometimes called 'inherent moisture'), but water from decomposition reactions is.

There are standards and locally followed practices for determining volatile matter in coals. Speight [14] states that in general tests are in the range 875–1050°C. The procedure is as follows. A lidded porcelain crucible containing a known weight of the coal, specified by whatever procedure or standard is being followed, is placed in an electric furnace at a set temperature which will be in the range given above. Safety measures apply. On commencement of decomposition of the volatiles, the air in the crucible is driven off and an inert atmosphere is created and sustained by evolution of the volatile matter which includes heavy components such as tars. After a pre-determined time in the furnace the crucible is removed, allowed to cool and weighed. Clearly the volatile matter can be determined by subtraction, and can be expressed on an ash-free basis in the way described in the box. First note that the ash content will have been determined separately. Volatile matter, moisture content, ash content and 'fixed carbon' when determined for a particular coal constitute 'proximate analysis'. 'Fixed carbon' has been put in inverted commas because it is not composed solely of carbon but contains an appreciable amount of hydrogen. It is what remains of the coal when the volatiles have been removed by heating. These terms make an appearance in later parts of this book including section 6.9.3.

A lignite having been dried in a desiccator is subjected to a volatile matter determination in a quantity of 10 g. The lignite is 5% ash. If the weight of solid remaining after the test is 5.2 g determine the volatile content on an ash-free basis.

The fact that the coal has been dried means that in this example 'ash-free basis' also means 'dry, ash-free basis'.[3] Loss of 4.8 g of coal substance was from:

(10 × 0.95) g organic coal substance = 9.5 g

So the volatile matter content is (4.8/9.5) × 100% = 51% ash-free basis.

Table 1.3 gives volatile matter values for three lignites and three black coals. They have intentionally been taken from the more recent literature.

Table 1.3 Volatile matter determinations for brown and black coals.

COAL	VOLATILE MATTER CONTENT
A Victorian brown coal [23]	49.4% dry basis
A North Dakota lignite [24]	47.9% dry basis
An Indonesian lignite [25]	37.1%
An Indian bituminous coal [26]	23% dry, ash-free basis. A 'medium volatile bituminous' according to the classification in [27]
Anthracite from Vietnam [28]	6.4%
Anthracite from China [29]	< 10%

It is clear from Table 1.3 that volatile matter declines very sharply with progression along the coalification series. One consequence of this relates to utilisation. A particle of pulverised fuel on entering the hot environment of a furnace will release its volatiles rapidly, and for a lignite these will provide at least as great a proportion of the total calorific value as the residual char.

1.4 Moisture and its removal

It was emphasised in section 1.3.1 that lignites as mined are very high in moisture. Water removal is therefore required in most applications. Methods for moisture removal from lignites have recently been reviewed [30]. What follows draws on the review.

Moisture removal can be evaporative or non-evaporative. In the former the heat of vaporisation of the water is provided. An obvious and widely practised means of evaporative drying at power stations is use of heat from the flue gases. Redirection of the flue gases requires fans and the like with a not insignificant electricity requirement which subtracts from the output of the power station. Sometimes heat for drying is

3 There have been criticisms of reliance on a conventional desiccator totally to dry a coal sample. The alternative is heating at about 105°C which, even if it is done under nitrogen, can cause weight loss by chemical reaction.

obtained not from the post-combustion gas but from a separate natural gas burner. This applies in particular to dryers of the rotary type (although use of flue gases with these is possible).

Non-evaporative moisture removal methods include mechanical thermal expression (MTE), which works by application of mechanical pressure. With mechanical methods of drying the lignite is likely to have been heated to soften it before application of pressure. Typical conditions are 150–200°C and 6 MPa of applied pressure. Another non-evaporative approach is hydrothermal dewatering (HTD). In this the lignite is heated to a temperature in the approximate range 250–310°C. Resultant shrinkage forces the water, in liquid form, out of the lignite particles. There is a patented process for low-rank coal drying called K-Fuel®. By heat and pressure this process removes water, but it does more than just remove water. It removes mercury, emission of which with flue gases after combustion is harmful. Conditions are sufficient for the structure of the coal to be changed to the enhancement of combustion performance.

Changes to the coal structure also occur with a continuous form of HTD which is at the demonstration stage in Australia amongst other countries. It removes 60% of the moisture as well as bringing about decarboxylation. There is also hot water drying (HWD), originating in North Dakota. In this the coal is heated to about 240°C, and there is saturation pressure of water vapour. Decarboxylation occurs, and carbon dioxide release breaks open the pores and in exiting the particle takes water with it. The presence of water vapour prevents ejection of tars and oils which are immiscible with water. That would of course affect the calorific value. Application of MTE and HTD to a particular brown coal from Australia is discussed in section 10.2.2.

There is also the screw conveyor dryer as a means of drying lignite. In this, as coal is conveyed it receives heat across a metal boundary. This process, which is of course in the evaporative category, is currently only exploratory, as is the use of microwaves in lignite drying. This is not so for fluidised bed drying, long a way of removing moisture from coals of any rank. There is R&D into fluidised bed drying of lignites, including bed pulsation at around 40 Hz to invigorate processes at the particle–gas interface. Use of superheated steam in drying, by direct contact with the lignite, also takes place, conditions being such that there is no return of the steam to saturated conditions. It has been found that increased grindability is obtained as a bonus when superheated steam is used in drying.

1.5 Concluding remarks

This introductory chapter has dealt with physical properties such as calorific value and hardness as well as with drying. It will be followed by one in which the petrographic side of the subject is introduced. All of these are required for a holistic coverage of the topic.

1.6 References

[1] *International Energy Outlook 2013 with Projections to 2040*, United States Energy Information Administration (2013)

[2] http://www.worldcoal.org/coal/what-is-coal/

[3] https://lh6.googleusercontent.com/-uYR5IS2tPy0/TWp2wlStO1I/AAAAAAAAALY/KeR-rKl6R9EQ/s1600/coalification.jpg

[4] http://geology.com/rocks/coal.shtml

[5] http://www.globalccsinstitute.com/insights/authors/dennisvanpuyvelde/2013/07/05/reducing-emissions-brown-coal

[6] *DryFining™ Fuel Enhancement Process*, Great River Energy (2014)

[7] Rattanakawin C., Tara W. 'Characteristics of Mae Moh lignite: Hardgrove grindability index and approximate work index', *Songklanakarin Journal of Science and Technology* 34: 103–107 (2012)

[8] http://www.cba-ssd.com/Applications/knowledgeBase/PRBcoal/PRBcoalProperty.htm

[9] http://pubs.usgs.gov/fs/fs004-02/fs004-02.html

[10] Jeffrey L.S. 'Characterization of the coal resources of South Africa', *Journal of the South African Institute of Mining and Metallurgy* February: 95–102 (2005)

[11] http://toolkit.pops.int/Publish/Annexes/A_28_Annex28.html

[12] Pourhoseini S.H., Moghiman M. 'Effect of pulverized anthracite coal particles injection on thermal and radiative characteristics of natural gas flame: an experimental study', *Fuel* 140: 44–49 (2015)

[13] Varol A.M., Atimtay A.T. 'Combustion of olive cake and coal in a bubbling fluidized bed with secondary air injection', *Fuel* 86: 1430–1438 (2007)

[14] Speight J.G. *The Chemistry and Technology of Coal*, 3rd edition, CRC Press (2012)

[15] http://www.sgs.com/Mining/Analytical-Services/Coal-and-Coke/Hardgrove-Grindability-Index-HGI.aspx

[16] http://www.energy.psu.edu/HGI/index.html

[17] Schobert H.H. *Lignites of North America*, Elsevier (1995)

[18] http://www.gilgames.com.au/coalspec.html

[19] Sherry A., Beck J.S., Cruddace A.E. *Modern Power Station Practice*, Elsevier (2014)

[20] Rayaprolu K. *Boilers for Power and Process*, CRC Press (2009)

[21] *Coal Hardgrove Grindability Index Certificate of Analysis Standard Reference Material 002* (2013)

[22] Özbayoglu G., Özbayoglu M.E. 'A new approach for the prediction of ash fusion temperatures: a case study using Turkish lignites', *Fuel* 85: 545–552 (2006)

[23] Mollah M.M., Jackson W.R., Marshall M., Chaffee A.L. 'An attempt to produce blast furnace coke from Victorian brown coal', *Fuel* 148: 104–111 (2015)

[24] Mangena S.J., Bunt J.R., Waanders F.B., Baker G. 'Identification of reaction zones in a commercial Sasol-Lurgi fixed bed dry bottom gasifier operating on North Dakota lignite', *Fuel* 90: 167–173 (2011)

[25] Zhao H., Yu J., Liu J., Tahmasebi A. 'Experimental study on the self-heating characteristics of Indonesian lignite during low temperature oxidation', *Fuel* 150: 55–63 (2015)

[26] Awasthi S., Awasthi K., Ghosh A.K., Srivastava S.K., Srivastava O.N. 'Formation of single and multi-walled carbon nanotubes and graphene from Indian bituminous coal', *Fuel* 147: 35–42 (2015)

[27] Thomas L. *Coal Geology*, John Wiley (2002)

[28] Muraoka M., Tominaga H., Nagai M. 'Iron addition to Vietnam anthracite coal and its nitrogen doping as a PEFC non-platinum cathode catalyst', *Fuel* 102: 359–365 (2012)

[29] http://www.fsanthracite.com/enarticle.php?Cid=11&TypeID=2

[30] Nikolopoulos N., Violidakis I., Karampinis E., Agraniotis M., Bergins C., Grammelis P., Kakaras E. 'Report on comparison among current industrial scale lignite drying technologies (A critical review of current technologies)', *Fuel* 155: 86–114 (2015)

CHAPTER 2
PETROGRAPHY

2.1 Introduction

It is to Marie Carmichael Stopes that we owe the basis of the science of coal petrography, although her place in history probably has more to do with her activity in another sphere altogether. She coined the term 'maceral', the organic analogue of 'mineral'. The concept has been fruitful in enabling coals to be characterised on a petrographic basis but there is one important distinction [1]: a maceral, unlike a mineral, does not have a precise chemical composition. To 'macerate' something is to soften it by steeping it in water, which is what happens at the initiation of coalification when vegetation becomes immersed in a swamp, and that is the origin of the term. Further introductory discussion follows.

2.2 Maceral groups and examples from lignite usage

Table 2.1 gives some details of coal constituents and below follows a discussion of how they apply to lignites in particular. The basic form of the table is taken from [2], and will be expanded upon by reference to other sources. The terms in the left-hand column refer to groups of macerals as explained in [3].

Table 2.1 Outline of coal constituents on a petrographic basis.

CONSTITUENT	DESCRIPTION	COMMENTS
Vitrinite	Occurs as bands	Assessed in terms of reflectance
Liptinite	Resinous	Higher content of hydrogen than inertinite (see below)
Inertinite	Charcoal-like	Unresponsive to thermal decomposition
Minerals and inorganics	Including clay (mineral) and calcium ions (inorganic)	Ash-forming constituents

Sub-grouping of macerals is carried out and frequently reviewed and modified, and terminology well exceeds in complexity and precision the summary given here. Even so, the level of petrography in this section will enable a reader to grasp the petrography content of literature on such topics as combustion, gasification, pyrolysis and liquefaction of lignites and provide him or her with the wherewithal if necessary to go more deeply into the subject.

'Vitrinite reflectance' measurements are routine in laboratories concerned with coals of any rank. Their determination involves microscopic examination under plane polarised light [4] and is quite simply the percentage of the incident light that is reflected by the vitrinite onto which the plane polarised light is focused. Liptinite can also be examined by reflectance. A standard for vitrinite reflectance measurements is mentioned in section 24.1. Vitrinite is very often the most plentiful of the maceral groups in a particular coal. Inertinite is the least reflective of the three groups. It is noted in [5] that whereas in a particular coal liptinite has a higher hydrogen content than inertinite, that of vitrinite can be higher or lower than that of liptinite. Minerals are distinct from yet analogous to macerals as explained in the opening paragraph. Their relevance to coal utilisation is the negative effect of their creating ash. Extraction of groups of macerals from a coal, for example by centrifuging, is routine laboratory practice (e.g. [6]). 'Maceral concentrates' for laboratory use are obtained in this way.

A further important concept is 'gelification', expressible as the gelification index (GI) which is calculable from amounts of particular macerals. The GI of a number of the lignites which feature in this book have been determined and are given in their due place. In its simplest form the index is [7]:

huminite macerals (% volume) ÷ inertinite macerals (% volume)

although this is often modified for particular deposits of lignite to incorporate specific macerals and not groups as in the above expression. That is so of the two examples of GI calculations – one for a Bulgarian lignite and one for an Indian lignite – later in this book.

Gelification involves loss of woody texture in going from peat to lignite, there being an intermediate state which is colloidal.

2.3 Examples of applications to lignites

Selected literature is drawn on to exemplify the application of the principles expressed in quite a basic way in the previous section to utilisation. Tabular presentation (Table 2.2) followed by comments is believed to be the most suitable approach. Obviously later parts of the book will be anticipated, for example where petrographic composition is correlated with liquefaction. Cross-references will be made as necessary.

Table 2.2 Research reports on lignites in which petrography features.

REFERENCE	DETAILS
[8]	Victorian brown coal examined for interaction with a hydrogen-donating solvent (tetralin) and effects on the huminite group of macerals examined
[11]	Lignites from North Dakota, Texas and Saskatchewan (two) examined for potential for conversion to liquid fuels
[12]	Vitrinite reflectances for a range of Chinese lignites
[14]	A 'Beulah Zap' lignite[4], as well as other coals of different rank, examined for combustion behaviour in a thermal analysis system with mass-spectrometric analysis
[15]	Three Turkish lignites separated into maceral groups
[17]	See comments in the main text
[19]	See comments in the main text

With reference to the first row of the table, the term 'huminite' needs to be clarified, and is an example of variation in nomenclature. 'Huminite' is described in [9] as being equivalent to vitrinite, though applicable only to brown coals/lignites. Reference [10] goes as far as to say that 'huminite is a synonym for vitrinite in lower rank coals' and invokes an Australian standard in so doing (see also sections 4.2.2, 18.2 and 24.1). That being said, the term 'huminite' is sometimes applied to sub-bituminous coals (see section 21.2). This is an example of the inconsistency which arises from maceral classification which is, to some degree, arbitrary.

In [8] a plasticising effect on the huminite macerals in gelified material is noted. There was no such effect for huminite in ungelified part. Indeed, the point is made in [8] that the effect of the tetralin on the huminite macerals is similar to that on vitrinite macerals when tetralin is applied to higher rank coals. In [11] (following row) vitrinite reflectance across four lignites examined was in the range 0.24–0.36%. In the same piece of work ten sub-bituminous coals from Alberta were included, vitrinite reflectances for which were in the range 0.33–0.51. Moving on to the following row, the vitrinite reflectances for the four Chinese lignites were in a narrow range: 0.41–0.46%. There are different approaches to reporting vitrinite reflectance results, and they include maximum reflectance and random reflectance, multiple measurements having been made in either case [13]. The authors of [12] used random reflectance.

In considering [14] we first return to the statement above that vitrinite is the most abundant maceral group in any one particular lignite. The maceral analysis for the Beulah lignite [14] is vitrinite 80%, liptinite 5% and inertinite 15% volume basis. The vitrinite reflectance is 0.18%. The authors of [14] were interested in CO/CO_2 ratios from the burning of the coals and the maceral concentrates in an atmosphere of 20% oxygen balance helium. (Use of air is of course precluded by the fact that carbon monoxide and

4 As is also pointed out in Chapter 4, Beulah Zap is a zone of the nearby Freedom Mine.

nitrogen have the same molar masses so cannot be distinguished on a mass spectrum.) The ratio of peaks for mass number 28 (CO) to mass number 44 (CO_2) for the Beulah lignite was 0.22, lower than the value for some of the higher rank coals.

In summarising reference [15] we note that it uses the term 'exinite' instead of 'liptinite', the former being an obsolescent term (reference [15] is from 35 years ago) synonymous with the latter. In this study lignites from five sources within Turkey were subjected to maceral analysis. The sources were Can, Seyitomer and Tuncbilek. The resolution was into the three maceral groups and introduces a point not previously made in this chapter: that coals from the same source having different particle densities will, as would intuitively be expected, have different maceral compositions. This is however highly apparent for the Tuncbilek lignite, where the less dense particle had 88.9% vitrinite and the denser one 54.5%. This becomes less difficult to understand if it is compared with classical 'float-sink' and 'heavy media' procedures for preparing coal for utilisation [16], an approach the authors of [15] themselves take. The principle of that is sinking of the mineral-rich particles and floating of the particles low in minerals to give floats, sinks and middlings. The Tuncbilek lignite which had become so depleted in vitrinite was 43.1% in ash compared with 7.8% for the less dense form. Otherwise the trends for the three Turkish lignites were as one would expect, with vitrinite (apart from the anomalous example described above) in the range 82.2–88.9%, exinite in the range 1.1–5.5% and inertinite in the range 1.3–8.8%.

In [17] it is asserted that in lignites inorganic elements are present as dissolved salts in the macerals and that this is less so for higher rank coals. There is a point of contact between this and a contribution to the literature from 25 years earlier when it was argued that for brown coals 'minerals' and 'inorganics' should be distinguished in analysis reporting [18], a point which was touched on in Chapter 1. In [19] the point is made, by citation of independent work, that when isolated macerals are examined for pyrolysis they do not always perform equivalently to the same macerals when in co-existence in the coal. This obviously indicates interaction between them in pyrolysis.

2.4 Lithotypes

In newly created surface in a brown coal deposit from which most of the water has evaporated, bands of quite perceptibly different shades of brown become apparent and these are called lithotypes. They are given descriptors accordingly, such as 'dark', 'pale', 'medium-dark' and so on. Their existence is due partly to the degree of gelification and also to variations in the structure and composition of the plant life at the outset of maceration. Classification according to colour and other macroscopic properties is frequently applied to lignites [9] and is quantified as a 'colour index' determinable by microscopic means. The older classification due to Stopes is on the basis of visible 'bands' termed vitrain, clarain, durain and fusain and has a petrographic basis; there are variants on these terms. This is applied to higher rank coals but also sometimes to lower rank ones. For example, such information for the Beulah Zap mine is given in [20]. Petrographic information based on lithotype features in later parts of the book including sections 4.2.4 and 21.2.

2.5 Subsea lignite

Coal deposits often extend beyond coastal boundaries. As long ago as 1860 there was mining of coal under the sea off Japan, an extension of the Takashima coal mine [21], and there has been subsea coal mining much more recently than that in many places including England and Nova Scotia. It is well known that the Kent coalfield in south-east England extends beyond the Thames estuary to the English Channel, and also that there is coal under Sydney Harbour.

During a subsea drilling operation for oil in the 1970s, a lignite deposit was discovered beneath the Indian Ocean [22] at Ninety-East Ridge. There is lignite beneath the Marmara Sea in Asia Minor [23]. There is subsea lignite at Sozopol Bay, off Bulgaria. These few examples could of course be multiplied many times.

2.6 Autochthonous and allochthonous lignites

The former occur at the site of the original plants, whilst the latter occur at sites to which the plant material has moved by the agency of water or of wind [24]. Although all of the lignites in current use in the USA are autochthonous [25], Arkansas lignite (see section 6.7) is allochthonous [20]. Where the plant debris without being moved away from the deposition site has been moved internally by circulating water, the term hypautochthonous applies. These terms are used at later stages of the book in relation to particular lignites; see in particular section 5.10.

2.7 Concluding remarks

The above, drawing on literature accounts, will have helped a reader of this book in two ways: by providing an introduction to nomenclature and by introducing him or her to the role of macerals – the fundamental unit in coal petrography – in several operations involving lignites all of which feature later in the book. Terminology is sometimes uncertain, as is discussed more fully in Chapter 21. It is emphasised again that the terms presented previously – vitrinite (huminite in lower rank coals), liptinite (a.k.a. exinite) and inertinite –represent only *groups* of macerals. An authoritative breakdown of groups into individual macerals is in [26], which is a publication of the United States Geological Survey and which is drawn on in later parts of this book. It should however be applied with an awareness that some maceral names are restricted to coal of certain rank. The huminite/vitrinite identity previously described in this paragraph is an example of that. Other geological terms are introduced as necessary throughout the book. Vitrinite reflectance has received some coverage in this chapter, and occurs in several later chapters. The in some ways complementary method of microfluorescence photometry is deferred until Chapter 21.

2.8 References

[1] Jones J.C. 'The nature of macerals', *Fuel* 89: 1743 (2010)

[2] http://www.coalmarketinginfo.com/advanced-coal-science/

[3] http://www.coaltech.com.au/InfluenceofCoalProperties.html

[4] http://www.astm.org/Standards/D7708.htm

[5] http://www.coaltech.com.au/InfluenceofCoalProperties.html

[6] http://pubs.acs.org/doi/abs/10.1021/ef00034a005?journalCode=enfuem

[7] Davis R.C., Noon S.W., Harrington J.W. 'The petroleum potential of Tertiary coals from Western Indonesia: relationship to mire type and sequence stratigraphic setting', *International Journal of Coal Geology* 70: 35–52 (2007)

[8] Shibaoka M. 'Behaviour of huminite macerals from Victorian brown coal in tetralin in autoclaves at temperatures of 300–380°C', *Fuel* 61: 265–270 (1980)

[9] Flores R.M. *Coal and Coalbed Gas: Fuelling the Future*, Newnes (2013)

[10] Li C-Z. *Advances in the Science of Victorian Brown Coal*, Elsevier (2004)

[11] Parkash S., Du Plessis M.P., Cameron A.R., Kalkreuth W.D. 'Petrography of low-rank coals with reference to liquefaction potential', *International Journal of Coal Geology* 4: 209–234 (1984)

[12] Wilson M.A., Vassallo A.M., Liu Y.L., Pang L.S.K. 'High resolution solid state nuclear magnetic resonance spectroscopy of Chinese maceral concentrates', *Fuel* 69: 931–933 (1990)

[13] http://www.coalpetrography.com/blog1/services/vitrinite-reflectance/

[14] Varey J.E., Hindmarsh C.J., Thomas M.K. 'The detection of reactive intermediates in the combustion and pyrolysis of coals, chars and macerals', *Fuel* 75: 164–176 (1996)

[15] Dogru A.R., Gökcen S.L. 'Determination of dominant macerals from some Turkish Tertiary lignites by means of heavy-liquid separation: a case study', *Fuel* 59: 355–356 (1980)

[16] http://www.sgs.com/Mining/Production-and-Plant-Services/In-Plant-and-Operational-Support/Preparation-Plant-Services/Float-Sink-and-Washability-Testing.aspx

[17] Teixeira P., Lopes H., Gulyurtlu I., Lapa N. 'Use of chemical fractionation to understand partitioning of biomass ash constituents during co-firing in fluidized bed combustion', *Fuel* 101: 215–227 (2012)

[18] Kiss L.T., King T.N. 'The expression of results of coal analysis: the case for brown coals', *Fuel* 56: 340–341 (1977)

[19] Bragato M., Joshi K., Carlson J.B., Tenório J.S.S., Levendis Y.A. 'Combustion of coal, bagasse and blends thereof. Part I: Emissions from batch combustion of fixed beds of fuels', *Fuel* 96: 43–50 (2012)

[20] Schobert H.H. *Lignites of North America*, Elsevier (1995)

[21] McKelvey V.E. *Subsea Mineral Resources*, United States Geological Survey (1986)

[22] http://gsabulletin.gsapubs.org/content/85/8/1219.abstract

[23] Okay A.I., Kaslilar-Ozcan A., Imren C., Boztepe-Guney A., Demirbag E., Kuscu I. 'Ac-

tive faults and evolving strike-slip basins in the Marmara Sea, northwest Turkey: a multichannel seismic reflection study', *Tectonophysics* 321: 189–218 (2000)

[24] O'Keefe J.M.K., Bechtel A., Christanis K., Dai S., DiMichele W.A., Eble C.F., Esterle J.S., Mastalerz M., Raymond A.L., Valentim B.V., Nicola Wagner N.J., Ward C.R., Hower J.C. 'On the fundamental difference between coal rank and coal type', *International Journal of Coal Geology* 118: 58–87 (2013)

[25] Cross A.T., Phillips T.L. 'Coal-forming plants through time in North America', *International Journal of Coal Geology* 16: 1–46 (1990)

[26] http://energy.usgs.gov/Coal/OrganicPetrology/PhotomicrographAtlas/OPTICCoalMaceralClassification.aspx

CHAPTER 3
LIGNITE/BROWN COAL IN PRE-INDUSTRIAL TIMES AND EARLY IN THE INDUSTRIAL ERA

3.1 Examples

Table 3.1 gives examples of lignite coal discovery and usage over a period encompassing AD 0; comments on the table follow.

Table 3.1 Scenes of early discovery and usage of lignites.

LOCATION AND REFERENCE	DETAILS
Westphalia, Germany [1]	Brown coal dug out for heating and cooking in Roman times. Scene of the current 'Tagebau Hambach mine'
North America [3]	Archaeological site in Texas which indicates burning of lignite by the Clovis peoples, possibly interchangeably with firewood
Louisiana Territory[5], USA [5]	Discovery of a formation comprising 'many horizontal strata of carbonated wood'
North Dakota[6], USA [7]	Small-scale use of lignite by indigenous Americans 'centuries before the settling [by whites] of North Dakota [in c.1860]'
Amsdorf, Germany [8]	Lignite mining from 1691
Roman Province of Germania Inferior, close to modern Cologne [9]	A record dated AD58 of an accidental fire initiated at a lignite deposit. Heavy damage to crops and structures
Silesia [10]	Lignite mining in the 17th century
The Urals [11]	Lignite mines extensively worked by the mid 18th century
Cannanore (a.k.a. Kannur), south-western India [14]	Discovery of lignite in 1830, the first such discovery in the whole of India
Ereğli, Black Sea coast [15]	A sample of lignite from Ereğli taken to Istanbul for evaluation. No immediate follow-up
Grevenbroich, Germany [17]	Discovery of lignite in 1858

With reference to the deposit in Westphalia (row 1), we note [2] that Tagebau means 'open cast mining'. (Correspondingly, 'Untertagebau' means 'underground mining'

5 A location which was later in Idaho Territory, then in Montana Territory then in the State of Montana.
6 Once part of the Louisiana Purchase.

[2].) Use of the deposit in modern times dates only from 1978. This was at least in part due to opposition to the destruction of settlements necessitated by development of the deposit. The Clovis peoples (next row) are believed to have lived ≈ 15 000 years ago [4], taking the timeline of this discussion into prehistory. The next row is concerned with an event in 1805, by which time the industrial era had begun though the events being described were certainly not in an industrial setting. In April that year Meriwether Lewis and William Clark (both US Army, on an expedition commissioned by President Jefferson) reported the discovery of what they termed carbonated wood as noted and they conducted a simple 'burn test' on the substance. Later investigation confirmed that 'carbonated wood' was classifiable as lignite according to criteria which at the time applied. The names of Lewis and Clark are perpetuated in many ways, their contribution to US cartography having been major. One such 'memorial' is the Lewis and Clark Power Station in Montana [6]. This uses lignite from the Savage mine, which straddles the border between Montana and North Dakota. The mine has an output of 0.35 megatonnes of lignite per year not all of which goes to the Lewis and Clark Power Station.

Reference [7] informs a reader that the earliest white settlers in North Dakota were shown the lignite deposits by the indigenous Americans who had previously made use of them. At Amsdorf (following row) lignite provided the heat needed for salt production. Salt had an important role in community health at that time, being used more to protect foodstuffs from microbial contamination than to flavour them. Lignite production at Amsdorf has been uninterrupted from the date given in the table to the present time and there is now a major electric power utility there. We are informed in reference [9] that by the time of the discovery lignite was seen both as an asset and as a hazard by reason respectively of its use as a fuel and its propensity to fire. Silesia (following row) was not unique in replacing wood with lignite as a general-purpose source of heat at that time, there being such substitution, for example, in central Germany in the early 1670s [10]. To the production of lignite in the Urals in the 18th century noted in the following row can be added other examples from countries which we now refer to as the Former Soviet Union. Highly interestingly [12], the Donets Coal Basin, part of which is in what is now the Ukraine and the remainder in Russia, contains coal across almost the whole range of rank from lignite to anthracite. Coal was discovered there in 1721. The lignite is at depths of about 600 m, the hard coal at depths up to about three times that [13]. At the time of the discovery of lignite at Cannanore (following row), India had not been partitioned so contained what later became Pakistan, where there is an abundance of lignite. The year 1830 is surprisingly late for the initial discovery of lignite in India; there was black coal production in India by 1775. For the last 60 years coal winning at Cannanore has been in the hands of the Neyveli Lignite Corporation.

In its bed-moist state lignite from Cannanore is about 50% moisture [16]. Fairly obviously the lumps of lignite shown in the illustration have been air dried. This is a suitable point at which to discuss the meaning of 'air dried' more generally. Sometimes it is identified with phase equilibrium between the coal's moisture content and the humidity of the atmosphere in which the coal has been placed. If the coal water was entirely in pores, acting as simple receptacles for the water, that would mean an atmosphere having the saturated vapour pressure of water at the temperature of the coal and its environment. The realism of this raises questions in the mind of a coal analyst, but there is in any case a further fundamental point to be made. Not all of the moisture content is in the pores. Some is imbibed in the cross-linked structure of the coal in a process analogous to being dissolved. Atmospheric water in equilibrium with the imbibed water would have a pressure lower than the saturation value and would be impossible to estimate. So the matter of the moisture content of a coal when 'air dried' has this conceptual complication. The matter of imbibition by coal of an introduced chemical agent will feature later in the book when the storage of sequestrated carbon dioxide in disused mines is discussed.

Returning to Table 3.1, the discouraging response to attempts to introduce Ereğli lignite in 1822 (following row) were followed seven years later [15] by a more positive outcome, and production began in 1835. Ereğli is one of a number of coal mines in the province of Zonguldak. Some of the others are productive of hard coal. Throughout the period of coal production in the region, the coastal location has made for ease of export. More information on the deposit at Grevenbroich (final row) follows in the next chapter.

3.2 Concluding remarks

The applications of lignite in this chapter pre-date oil availability and usage, though in some cases not by much. In the following chapters applications of lignites, in particular electricity generation, are discussed. Concurrently with its production from lignites and other ranks of coal, electricity has, over the period of interest, been produced with oil or natural gas as fuel. There will be coverage in a later chapter of the conversion of one to the other, that is, of the production of liquid fuels from lignites. A series of chapters on electricity generation from lignite in recent times now follows. Sometimes when such generation for a particular country is discussed there is digression into parts of the country having lignite without using it to make electricity.

3.3 References

[1] http://www.earthisland.org/journal/index.php/elist/eListRead/germany_to_expand_brown_coal_mines/

[2] http://dictzone.com/german-english-dictionary/tagebau

[3] http://www.texasbeyondhistory.net/faq/

[4] http://www.crystalinks.com/clovis.html

[5] http://www.lewis-clark.org/article/1570

[6] https://www.lignite.com/mines-plants/power-plants/lewis-clark-station/

[7] https://books.google.com.au/books?id=_ZzUGlayhuYC&pg=PA410&lpg=PA410&dq=north+dakota++indigenous+americans+lignite&source=bl&ots=bt-mNmMih6&sig=x-eq2QbmkvlqGZLzdI4L3uq9wEM4&hl=en&sa=X&ei=XHfRVIaTN4XxmAWhnIL-gAQ&ved=0CC0Q6AEwAw#v=onepage&q=north%20dakota%20%20indigenous%20americans%20lignite&f=false

[8] http://www.wachs-und-mehr.de/index.php/en/unternehmen/historie

[9] http://www.google.com.au/url?sa=t&rct=j&q=&esrc=s&-source=web&cd=5&ved=0CDQQFjAE&url=http%3A%2F%2Fwww.bund-nrw.de%2Ffileadmin%2Fbundgruppen%2Fbcmslvnrw%2FPDF_Dateien%2FBraun-kohle%2F2013_10_12_Lignite_mining_in_the_rhineland_Garzweiler_II_web.pdf&ei=htnS-VI3sG8G0mwWW7IGoDQ&usg=AFQjCNHjNeHtil0iDpTnRT3ernEmMzA0vw

[10] http://www.geo-sounds.de/komposition

[11] Shoemaker M.W. *Russia and the Commonwealth of Independent States,* Rowman & Littlefield Education, Lanham, MD (2014)

[12] *Coal Mining in the Ukraine: Opportunities for Production and Investment in the Donetsk Coal Basin*, US Environmental Protection Agency (2001)

[13] http://encyclopedia2.thefreedictionary.com/Donets+Coal+Basin

[14] http://www.fundinguniverse.com/company-histories/neyveli-lignite-corporation-ltd-history/

[15] Guney M. 'Underground mining operations in Zonguldak coal mines' (1966). Accessible at: http://www.google.com.au/url?sa=t&rct=j&q=&esrc=s&-source=web&cd=5&ved=0CDcQFjAE&url=http%3A%2F%2Fwww.researchgate.net%2Fpublictopics.PublicPostFileLoader.html%3Fid%3D548dd629d4c118a8358b46ae-%26key%3Df07eae0c-9794-4c02-83f9-5c3f003ab1b5&ei=8EXUVIK_OebOmwX2hoD-IBQ&usg=AFQjCNFsm33g6-DGDdSt6y-1yBVLOAwBwQ

[16] http://www.slideshare.net/santhoshsachithanantham/neyveli-lignite-corporation-limited

[17] http://www.rwe.com/web/cms/en/60110/rwe-power-ag/locations/lignite/neurath-power-plant/

CHAPTER 4
ELECTRICITY GENERATION I – GERMANY

4.1 Background

Germany, both before and during the division into West and East and since its reunification, has been noted for lignite utilisation for power generation. In Germany lignite is called 'braunkohle' (see footnote 7). It is reported in [1] that 162 billion kW-hour of electricity was produced in Germany from braunkohle in 2013.

There was mine illumination using electricity from brown coal in Germany in 1884 [2]. We are informed in [3] that at the beginning of the 20th century the whole of the industrial base of central Germany relied on electricity made from brown coal. In [4] it is stated that two brown coal fired power stations in Germany came into being in response to wartime conditions. One was that at Zschornewitz and the other that at Knapsack (Goldenberg). The one at Zschornewitz came into operation on 14 December 1914, sixteen months into World War I. At that stage it had six steam turbines with a total output of electricity of 16 MW and was fuelled by brown coal from a local open cut [5]. It was at that time the largest lignite-fired electricity generating utility in the world. It was close to Leipzig, and therefore in the German Democratic Republic during the years of separation. Indeed, its closure was the result of general restructuring after the two zones of Germany were allowed to reunite [6]. The power station at Knapsack/Goldenberg near Cologne had entered operation the previous year. These introductions at the outset of World War I betokened a huge increase in power generation capacity in Germany at that time. At the beginning of the war, Germany produced more electricity than England, France and Italy combined [7] and obviously the ready availability of brown coal was a major factor in that. Harbke (see next section) is another example.

It is recorded in [1] that for the same year hard coal – of which there is an abundance in Germany additional to the braunkohle – produced 116.4 billion kW-hour of electricity. The respective contributions to the total are 25.8% and 19.7%, the balance being electricity generation with fuel oil. Selected German plants for power generation will be discussed in turn. The usual combustion technique for lignite in power generation is as pulverised fuel (p.f.), which comprises particles having been milled to a mean diameter of typically 50–100 μm. These are entrained in air and taken to a burner where the resulting flame in several important ways resembles a gaseous jet flame and can be analysed analogously to that. Fluidised beds also find applications to lignites and selected examples of these will feature later in the text.

4.2 German power stations using lignite

4.2.1 Introduction

Details of major generating activity from lignite in Germany are given in what follows. It is good for the specialist in lignite to be aware that there is also major power generation from locally mined black coal in Germany.

4.2.2 Offleben and Buschhaus power stations

These are both in Lower Saxony, and lignite-fired. Offleben commenced operations in 1954 replacing Harbke power station [8,9] which, on division, was positioned within the DDR. Offleben was closed down in 2002, because of depletion of the nearby lignite pit from which it drew its fuel. The story with Buschhaus is much more positive. A 390 MW utility commissioned in 1980, Buschhaus is in the Helmstedt lignite mining area [10]. The existence of 'light', 'medium light' and 'dark' lithotypes in lignite from Helmstedt has been noted [11] and correlated with maceral analysis. Liptinite and inertinite feature in the analysis, as do for example 'humodetrinite' and 'humocollinite' which are precursors to vitrinite and are sub-classifications of 'huminite'.

The performance of Buschhaus in 2012 was 2.3 TW-hour. There is use of supercritical steam, a full discussion of which is given in the next section.

4.2.3 Grevenbroich-Neurath power station

It was noted in the previous chapter that braunkohle was discovered at Grevenbroich in the 1850s. Even so, power generation did not commence there until 1972 [12], and the power station is known as Grevenbroich-Neurath (G-N). Its nameplate output is 2200 MW, although it is operated at this rating only for a very small proportion of the time. The facility is owned by Rheinisch-Westfälisches Elektrizitätswerk AG (RWE AG).

The Grevenbroich-Neurath power station sources its braunkohle from a number of mines in that part of Germany including that at Hambach. Braunkohle from these sources are called collectively 'Rhineland lignites'[7] or 'Rhenish lignites' [13] and range in moisture content, when bed-moist, from 50 to 60%.

Expansion at Hambach of the area over which mining occurs, currently 35 km^2, will be necessary if targets for generation of electricity are to be realised. It has been noted and commented upon [14] that this would involve the destruction of trees. G-N is however making its contribution to carbon dioxide emission reductions. The efficiency of the Rankine cycle, the basis of a steam turbine, can be raised in a number of ways, and the higher the efficiency the lower the carbon dioxide emission per unit energy of electricity yielded. The (ideally) isentropic change in a turbine from superheated steam to predominantly liquid water is the basis of the production of work, and from analysis of a Rankine cycle (e.g. [15]) the greater the extent of superheating the greater the efficiency of conversion of heat to work. This is well known and occurs widely in texts on the laws of thermodynamics (e.g. [15]).

A typical Rankine cycle at a power station will operate at up to about 35% conversion of heat to work. At G-N the most recently installed turbines are attaining 43% efficiency. This is by use of supercritical steam, an introduction to the thermodynamics of which forms an appendix to this chapter. The appendix concludes by explaining how and why the 'ultra-supercritical' denotes degree and does not have a scientifically

7 Note that the traditional term in German usage braunkohle has in many instances been changed to 'lignite'.

rigorous distinction from simply 'supercritical'. Use of supercritical steam at G-N necessitated turbine construction of materials which can withstand the conditions [16]. The carbon mitigation accruing from the increased Rankine efficiency at G-N can be estimated without difficulty, and such a working is shown in the box. Numerical values other than the efficiencies are the author's estimates. Readers can if they wish repeat the calculation with other values for these quantities.

We take it that the lignite as fired has a calorific value of 15 MJ kg^{-1} and that its percentage carbon in the air-dried state is 60%. The basis of the calculation will be electricity generation at 500 MW.

A 500 MW turbine in operation round-the-clock would, in 30 days' usage, produce an amount:

500×10^6 J s^{-1} $\times$ (30 $\times$ 24 $\times$ 3600) s = 1.30×10^{15} J

At a Rankine efficiency of 35% the heat required would be:

$(1.30 \times 10^{15}/0.35) = 3.71 \times 10^{15}$ J

At a Rankine efficiency of 43% the heat required would be:

$(1.30 \times 10^{15}/0.43) = 3.02 \times 10^{15}$ J

The saving in heat is then 0.69×10^{15} J

At a calorific value of 15×10^6 J kg^{-1} the fuel saving is:

$= 4.60 \times 10^7$ J or 46 000 tonnes

The reduction in carbon dioxide is then:

$0.6 \times 4.60 \times 10^4 \times (44/12)$ tonne = 100 000 tonnes approx. in a 30-day period.

At a 2015 carbon credit price of 18 Great British Pounds (GBP) per tonne [17], the financial saving is GBP 1.8 million over the 30-day period of generation at 500 MW.

That use of supercritical steam in the turbines at G-N yields major financial benefits is therefore clear. One would expect intuitively that the payback time on the introduction of (ultra-)supercritical steam would not be too long to make for viability.

4.2.4 Frimmersdorf power station

This power station began operations in 1953 which, we will note in passing, was one year before the sovereignty of Germany was restored after World War II. It is reasonable to suppose that the occupying allies had a role in setting it up. The power station replaced one at the same site which, with lignite as fuel, had commenced operation in 1926 [18] and produced at a rate of 26 MW.

The nameplate capacity is 1665 MW and there is some use of supercritical steam. In an RWE publication from a decade ago [20] it is reported that at two 300 MW steam turbines at Frimmersdorf paper sludge is co-fired with lignite. It is best to analyse this in fairly general terms, using estimates for the quantities required, rather than to restrict the analysis to Frimmersdorf. Such a working is shown in the box, and we first note that paper sludge, unlike lignite, is carbon-neutral. Only carbon dioxide from fossil fuel combustion causes the carbon dioxide level of the atmosphere to rise. To burn a carbon-neutral substance is to put the carbon back where it was before removal by plant life.

We assign calorific values to the respective fuels as follows:

Lignite 15 MJ kg^{-1}

Paper sludge 7 MJ kg^{-1}

At an efficiency of 35%, 600 MW of electricity requires:

600/0.35 MW of heat = 1700 MW of heat approx.

If it is desired to raise 1% of the heat from the sludge instead of from the lignite, 0.99 kg of lignite has to be co-fired with a quantity of the sludge capable of releasing as much heat as 0.01 kg of the lignite.

Assigning the lignite a carbon content of 40%, when 1665 MW [19] of heat are produced from the lignite alone the annual production of carbon dioxide is:

$0.4 \times (1665 \times 10^6 \text{ J s}^{-1}/15 \times 10^6 \text{ J kg}^{-1}) \times (44/12) \times (365 \times 24 \times 3600) \text{ s year}^{-1} \times 10^{-3}$ tonne kg^{-1}

– 5.134 million tonnes

When lignite-sludge is used as the fuel, the fossil fuel derived carbon dioxide is 0.99 of that, or 5.083 million tonnes for electricity generation at 1665 MW over one year.

The difference is 0.051 million tonnes with a value in carbon credit terms of $\approx$ GBP 0.93 million.

This working is for substitution of the carbon-neutral fuel for the lignite only to an extent of 1%. Much higher percentages can in fact be used. The calculation is taken a little further in the second box.

The amount of sludge thermally equivalent to 0.01 kg of lignite is:

$(0.01 \times 15/7)$ kg = 0.021 kg

So the co-firing requires a proportion of sludge:

$[0.021/(0.99 + 0.021)] = 0.021$ or 2.1%

Although the calculation was intended to be general, it will be noted that reference [20] informs us that the sludge was present at 1–2%. The figure expressed in GBP for the carbon credits accruing over a year's generation at 600 MW converts to $US 1.42 million. There is a return to co-combustion of lignite with other fuels more generally in Chapter 20.

Lignite for Frimmersdorf is from the Garzweiler opencast mine in the Rhineland, which also supplies other local power stations. Its contains up to 60% water in the bed-moist state, and its calorific value in that state is about 9 MJ kg^{-1} [21].

The overburden – matter requiring excavation and removal – at Garzweiler contains major amounts of pyrite (FeS_2, a very common mineral in coals) [22] and this is typical of low-rank coal deposits in the Rhineland. Overburden has to be dewatered before excavation and assembly as a pile. The quantity of pyrite in the overburden is a major factor. Pyrite oxidises in air to form sulphuric acid according to:

$$2FeS_2 + 7O_2 + 2H_2O \rightarrow 2FeSO_4 + 2H_2SO_4$$

and water pollution by the sulphuric acid, commonly referred to as acid mine drainage, ensues. In [22] proposals to control this at Garzweiler by neutralisation of the acid with limestone or with fly ash are made and tested. Water in contact with samples of overburden on a laboratory scale gave a pH reading of 3.4 without any neutralising reagent. The action of fly ash is due to its hydroxyl ions, and can be enhanced by bubbling in some carbon dioxide. Fly ash with carbon dioxide raised the pH to 6.4, and limestone alone raised it to 6.2. This theme continues in section 4.2.6 where mining at Jänschwalde is discussed.

Petrographic analysis of Rhineland brown coals has been on the basis of lithotype, and a coverage of this from the 1980s [23] makes considerable use of a term not introduced in section 2.3 of the present text: xylite, a term like so many others in the subject area used by Marie Stopes. The term refers to bands in the coal, and in [23] the presence of xylite in Rhineland brown coals is linked to the milieu of the original vegetation. A drier setting leads to less xylite than wetlands do.

4.2.5 Lippendorf power station

This is nine miles from Leipzig, therefore in East Germany over the period of division of Germany. It was built over the period 1964–68, therefore during the regime of the DDR. Modernisation took place in the late 20th century. Owned jointly by Vattenfall and Energie Baden-Wurttemberg AG [24], it uses lignite from the reserve at Vereinigtes Schleenhain and has a nameplate capacity of 16 000 MW. Vereinigtes Schleenhain lignite is 4.3% in sulphur. It has 52.4% volatiles (dry basis) and 11.6% ash (also dry basis) [25]. Yellow-brown bands have been observed in the coal seams [26].

The sulphur content of the coal used at Lippendorf is fairly high as noted above, and the ash content is also fairly high. It was noted in section 1.3.2 that brown coal for power generation is often burnt in the form of pulverised fuel (p.f.), and the ash from that is of particle size microns or lower ('fly ash'). At Lippendorf as in other such plants release of fly ash with the flue gases is prevented by the well-proven technique of electrostatic precipitation, which removes > 99% of the fly ash particles [27]. Desulphurisation is also performed at Lippendorf. Sulphur in any fuel goes, in stoichiometric yield, to sulphur dioxide in the flue gases. The technique used in its removal at Lippendorf is reaction with lime, achieving 95% removal of sulphur dioxide from the flue gases [27]. The reaction leads to gypsum, a saleable product [28].

4.2.6 Jänschwalde power station

Close to the border between Germany and Poland, Jänschwalde is also territorially within the former DDR. The plant, having been commissioned in 1988, was an enterprise of East Germany subsequently brought up to western standards in terms of emission control. The owner is Vattenfall. It now has a capacity of 3000 MW, from six 500 MW steam turbines. It uses lignite from mines including Jänschwalde and Cottbus-Nord. Having originally been in the DDR, Jänschwalde had to be retrofitted with flue gas desulphurisation equipment.

Jänschwalde is in the region of Germany known as Lusatia, and a review of the several lignite deposits in Lusatia arguably belongs to this part of the book. That at Vereinigtes Schleenhain has been discussed above at length. Others are listed in Table 4.1, comments on which follow below. The important matter of lignite mining dumps features *en passant* in this section.

Table 4.1 Lignite deposits in the Lusatia region of Germany.

LIGNITE RESERVE	DETAILS
Jänschwalde	Planned cessation of production in 2020 [29]
Cottbus-Nord	First producing in 1978. Future use only to 2018 [33]
Welzow-Süd	Production expected to continue at least until 2030 [34]
Nochten	Sulphur content being examined [37]. Lignite from this deposit currently in use, 0.7% sulphur in the bed-moist state [38]

Scheduled closure of the Jänschwalde lignite mine in 2020 is due to community resistance to expansion of the mine. The mine has featured in the research literature of the 21st century on a number of accounts including the following. Effects on the environment of the water resulting from mining at Jänschwalde are examined in [30]. The thrust of this work is prevention of damage to previously existing aquifers by water from mining activity, which is heavy in solutes. The 'dumping' theme continues into [31], in which pyrite oxidation features. It is recorded in [31] that at mines in Lusatia overburden depths are 44 ± 20 m. Some oxidation of pyrite to iron II sulphate occurs, and the 'turnover rate' – percentage of the pyrite so reacted – is relevant to the composition of mining dumps. In [30] the view is taken that 4–5% of the pyrite in a mine dump will be so oxidised. For Cottbus-Nord and Nochten (later rows of the table) a value of 6–7% is used. This is the 'turnover rate' over whatever time elapses between initiation of the oxidation of the pyrite in mining and the occasion of examination of the dump for possible dismantlement or whatever. The sale of gypsum from the Lippendorf power station featured in the previous section. In [32] the gypsum from Jänschwalde is compared for quality with that produced at a number of other German lignite-fired power stations.

Cottbus-Nord (next row) has over its 37 years of existence led, through expansion, to loss of residential land. A town called Lakoma was completely lost to the Cottbus-Nord mine. The site of the mine will be flooded on closure, becoming 'Cottbus East Lake' which, it is hoped, will have visual appeal and become a habitat for wildlife [32]. Welzow-Süd could have featured in Chapter 3, as there was limited lignite production there in the 19th century. What is now known as the Welzow-Süd mine came into being in 1959 [35]. A photograph of it can be seen in [36].

4.2.7 Boxberg power station

This came into service in the DDR in 1971 [39] and under that regime grew to a total output of 3520 MW. After reunification modernisation of the facility, directed at increasing turbine efficiency and reducing emissions, was implemented. A Vattenfall facility, it uses lignite from Lusatia. It uses supercritical steam, and turbine efficiencies of 42% have been reported.

Boxberg is an example of a power plant in the DDR needing modifications, including the installation of a desulphurisation facility, and Jänschwalde is referred to for the same reason in [40]. As at Lippendorf amongst many other power facilities, lime is the basis of the removal. At Boxberg the upgrading to supercritical steam has been carried out with newly developed metallic construction materials [41]. The Boxberg power station is alongside the Nochten mine.

The siting of power stations close to brown coal deposits is of course a widely followed practice at places including the Latrobe Valley. For the next German power station to be considered there will be a return to the Rhineland.

4.2.8 Niederaussem power plant [42]

An RWE facility, this entered service in 1963. Since then some units have been decommissioned and new ones have been introduced, and the current capacity is 2840 MW. Situated near Cologne, it uses Rhenish lignite. The matter of pulverising the lignite in readiness for firing has featured in previous sections of this chapter, and will be developed for Niederaussem. The power plant at Niederaussem has acquired a Loesche coal mill of type LM 28.3 D [43]. An illustration forms Figure 4.1.

Figure 4.1 Loesche LM 28.3 D coal mill. (courtesy Loesche Innovative Engineering, Düsseldorf, Germany)

Loesche manufacture a range of coal grinding mills the power input for which has a known dependence on the milling rate required. It ranges from 400 kW for a grinding rate of ≈ 50 tonnes per hour to 2400 kW for a grinding rate of ≈ 300 tonnes per hour [44]. The weight is of milled coal not of raw coal; the difference is due to moisture loss in milling. The LM 28.3 D coal mill requires 800 kW. The precise grinding rate depends upon the Hardgrove index and upon the size range of the coal as fed to the mill. Loesche themselves classify a coal as having 'easy grindability' if it has a Hardgrove index of 90 or higher [45]. This can be compared with values in Table 1.2.

There are nine units at the power plant at Niederaussem one of which, called K unit, uses supercritical steam and consequently achieves 43% efficiency [46]. Some of the lignite used at Niederaussem is subjected to fluidised bed drying [47]. This uses heat from the flue gases. It is pointed out in [47] that the heat of vaporisation of the water would otherwise have had to come from the coal, to the reduction of heat release and therefore of electricity output.

4.3 Further comments

A number of other lignite power stations in the reunified Germany could have featured above. These include the Schwarze Pumpe power station. Commissioned in

1996, it postdates some such facilities in Germany by a century (see also section 14.2). Schwarze Pumpe uses a grinding mill supplied by Loesche, the LM 35.3 D model. Reference to [44] reveals that this requires 1000 kW of power and delivers p.f. at a rate in a range overlapping with that of the LM 28.3 D used at Niederaussem. At Schwarze Pumpe there is a fluidised bed dryer for the lignite. This can receive 10 tonnes per hour of raw coal [48].

Schkopau uses 6 million tonnes per year of lignite from the deposit at Profen [40]. Profen is owned not by the operators of the power station but by the mining corporation MIBRAG [49] who themselves have plans to set up a power station in that part of Germany. There is a very long way to go with this undertaking by way of permits and no firm decision to build has yet been made. Even so, MIBRAG have released a stylised image of the Profen power station on their web pages (see [49]). MIBRAG also own a lignite mine at Vereinigtes Schleenhain in the same part of Germany. Lignite from this source is used for a combined heat and power (CHP) unit at Lippendorf.

In addition to the modern lignite-fired power plants identified so far in this section, there are many which have been taken out of service. These include Offleben power station which ceased activity in the early 2000s having been in service for about half a century.

Lignite therefore continues to be a major source of fuel for power generation in Germany. Germany once led in lignite utilisation, and combustion practices developed there were followed for example in Australia's Latrobe Valley, which has major coverage later in this book. It is hard to see how carbon mitigation should be more difficult for lignites in such usage than for hard coals. Both can be co-fired with biomass. Carbon dioxide sequestration methods can be applied with either. Perhaps the continued activity in lignite in Germany, not of course to the exclusion of 'renewables', follows from such musings. As power plants in Germany are being replaced there is an increasing move to supercritical steam which, it is argued in a later chapter of the book, is motivated by carbon mitigation as an alternative to carbon capture and storage (CCS).

4.4 Lignite in Austria

The neighbouring country Austria has lignite at places including the Köflach Voitsberg deposit which takes in the Oberdorf open cut mine [50]. The dominant lithotype is xylo-detritic coal which, in the petrographic terminology defined in [50], means dark rather than pale in colour and banded (see section 4.2.4 above) with some xylite evident. The calorific value ranges from 12 to 13 MJ kg^{-1} and the moisture content from 35 to 40%. Anticipating section 15.5.3, the lignite reserves of Austria contain major amounts of jet [51].

There was once lignite mining in Austria, and it began to decline in the 1960s [52]. There has been no production of any rank of coal in Austria since 2006. She currently imports all of her coal, and this includes some lignite. She imports some electricity from Germany.

4.5 References

[1] http://www.germanenergyblog.de/?p=15019

[2] *The Discovery of Electricity: Early Days of Brown Coal*, Powercor Australia (n.d.)

[3] http://www.erih.net/regional-routes/germany/saxony-anhalt/saxony-anhalt-detail/print.html

[4] Hughes T.H. *Networks of Power: Electrification in Western Society, 1880–1930*, JHU Press (1993)

[5] http://cosplaxy.com/index.php?lan=DE&key=1Ntzyi8fVdJBIaO8hqC+Iwdn8PGjukP9P7g-choL7Lyc

[6] http://www.industrielles-gartenreich.de/english/03_projekte/314_kolonie.htm

[7] Topik S.C., Wells A. *Global Markets Transformed*, Harvard University Press (2014)

[8] http://forum.skyscraperpage.com/showthread.php?t=186508

[9] http://www.divisionleap.com/pages/books/22990/photo-albums/photo-album-documenting-the-1922-renovation-of-the-harbke-power-plant

[10] http://www.euracoal.org/pages/layout1sp.php?idpage=72

[11] Winkler E. 'Organic geochemical investigations of brown coal lithotypes: a contribution to facies analysis of seam banding in the Helmstedt deposit', *Organic Geochemistry* 10: 617–624 (1986)

[12] http://www.rwe.com/web/cms/en/60110/rwe-power-ag/locations/lignite/neurath-power-plant/

[13] http://www.mining-technology.com/projects/rhineland/

[14] http://www.foeeurope.org/Forest-must-not-cleared-coal%20mine-say-green-groups-141112

[15] Jones J.C. *The Principles of Thermal Sciences and their Application to Engineering*, Whittles Publishing, Caithness and CRC Press, Boca Raton (2000)

[16] http://www.rwe.com/web/cms/en/12068/rwe-power-g/locations/lignite/kw-neurath-boa-2-3/

[17] http://www.theguardian.com/environment/2015/feb/24/european-carbon-emissions-trading-market-reform-set-for-2019

[18] http://cache.worldlibrary.net/articles/frimmersdorf_power_station

[19] http://www.rwe.com/web/cms/en/9002/rwe-power-ag/locations/lignite/frimmersdorf-power-plant/

[20] *Frimmersdorf and Neurath Power Plants: Electricity from Rhenish Lignite*, RWE Power (2005)

[21] *Lignite Mining in the Rhineland Garzweiler II*, Bund für Umwelt und Naturschutz Deutschland, Dusseldorf (2013)

[22] Obermann P., Wisotzky F. 'Acid mine groundwater in lignite overburden dumps and its prevention – the Rhineland lignite mining area (Germany)', *Ecological Engineering* 17: 115–123 (2001)

[23] Hagemann H.W., Wolf M. 'New interpretations of the facies of the Rhenish brown coal of

West Germany', *International Journal of Coal Geology* 7: 335–348 (1987)

[24] http://powerplants.vattenfall.com/powerplant/lippendorf

[25] Zhong S., Baitalow F., Nikrityuk P., Gutte H., Meyer B. 'The effect of particle size on the strength parameters of German brown coal and its chars', *Fuel* 125: 200–205 (2014)

[26] http://www.laop-consult.de/en/projects/evaluation_of_a_coal_seam_sequence_in_the_surface_mine_v_schleenhain/

[27] http://www.power-technology.com/projects/lippendorf/

[28] http://fam.de/english/News/newsarchive/news.20.html

[29] http://www.lausitzer-braunkohle.de/english.php

[30] Hoth N., Wagner S., Hafner F. 'Predictive modelling of dump water impact on the surroundings of the lignite dump site at Jänschwalde (Eastern Germany)', *Journal of Geochemical Exploration* 73: 113–121 (2001)

[31] Graupner B.J., Koch C., Prommer H., Werner F. 'Process oriented quantification of mine dump pollutant inventories on the large scale – the case of the lignite mining district Lusatia, Germany', *Journal of Geochemical Exploration* 112: 161–173 (2012)

[32] Hansen B.B., Kiil S., Johnsson J.E. 'Investigation of the gypsum quality at three full-scale wet flue gas desulphurisation plants', *Fuel* 90: 2965–2973 (2011)

[33] http://ejatlas.org/conflict/lignite-mining-cottbus-nord-lakoma-germany

[34] http://www.iba-see2010.de/en/projekte/projekt8

[35] http://www.erih.net/nc/countries/detail.html?user_erihobjects_pi2%5B-showUid%5D=16742

[36] http://www.bgr.bund.de/DE/Themen/Energie/Bilder/WeichbraunkohleWelzow_Bild1_g.jpg?__blob=normal&v=2

[37] http://www.laop-consult.de/en/projects/mapping_of_sulfur_content/

[38] http://www.infomine.com/news/newsletters/websites/editorials/suppliers/00

[39] http://powerplants.vattenfall.com/powerplant/boxberg

[40] Hansen U. 'Restructuring the East German energy system', *Energy Policy* 24: 553–562 (1996)

[41] http://www.modernpowersystems.com/features/featuresupercritical-pressure-power-plants-a-progress-report/

[42] http://www.rwe.com/web/cms/en/60132/rwe-power-ag/locations/lignite/niederaussem-power-plant/

[43] http://www.mining-technology.com/contractors/crushers/loesche/press6.html

[44] http://www.loesche.com/en/products/dry-grinding-plants/coal-for-cement/

[45] *Loesche Mills for Solid Fuels*, Loesche GmbH, Dusseldorf (2012)

[46] https://online.platts.com/PPS/P=m&s=1029337384756.1478827&e=1096495472414.2240023308785804128/?artnum=x2F004ShER07261656F2X4_1

[47] http://kraftwerkforschung.info/en/lignite-drying/

[48] http://www.bbs.bilfinger.com/en/innovation/lignite-drying/

[49] http://www.mibrag.de/index.php?id=3217#

[50] Kolcon I., Sachsenhofer R.F. 'Petrography, palynology and depositional environments of the early Miocene Oberdorf lignite seam (Styrian Basin, Austria)', *International Journal of Coal Geology* 41: 275–308 (1999)

[51] Bechtel A., Gratzer R., Sachsenhofer R.F. 'Chemical characteristics of Upper Cretaceous (Turonian) jet of the Gosau Group of Gams/Hieflau (Styria, Austria)', *International Journal of Coal Geology* 46: 27–49 (2001)

[52] Pedraza J.M. *Electrical Energy Generation in Europe*, Springer (2015)

APPENDIX TO CHAPTER 4
SUPERCRITICAL STEAM AND ITS APPLICATION TO STEAM TURBINES IN POWER GENERATION

Referring to the phase diagram for water in Figure A4.1, at a higher temperature than 374°C (647 K) and a higher pressure than 218 bar (22 MPa) water is in the supercritical state.

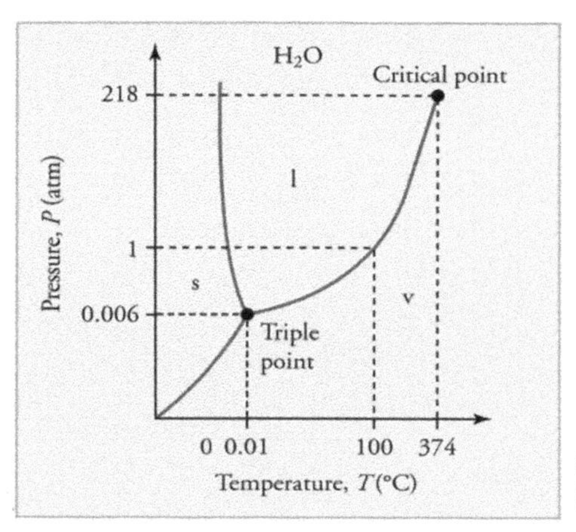

Figure A4.1 Phase diagram for water showing the supercritical region. Taken from [1].

A steam turbine using supercritical steam can be analysed by the Rankine cycle, which of course is the basis of turbine operation when saturated or superheated steam is used. Quite simply, in a cycle using supercritical steam the point of origin of the work-producing step represents steam at some point in the supercritical region of the diagram above. Supercritical steam is not, of course, restricted to power plants which use lignite as fuel. The term 'ultra-supercritical' denotes steam in the supercritical state distant from the critical point. As already pointed out, there is no fundamental difference in thermodynamics between supercritical and ultra-supercritical steam. An arbitrary definition is that supercritical steam above about 600°C is ultra-supercritical. The term 'highly supercritical', used in [2] in discussions of developments at German power stations generally, is probably preferable to 'ultra-supercritical steam' though much less frequently used. By analogy the term 'ultra-superheated steam', which one frequently encounters for example in accounts of gasification technologies, simply means steam having a high degree of superheating being a long way from saturation conditions. The point might legitimately be made that pulverised fuel combustion was starting to become a dated combustion technique until the introduction of supercritical steam stimulated a revival.

References in the Appendix

[1] http://cnx.org/resources/daad6e60f1896a0016876b4f0448456b/Figure_14_05_05.jpg

[2]http://www.modernpowersystems.com/features/featureultra-supercritical-chp-getting-more-competitive/

CHAPTER 5
ELECTRICITY GENERATION II – OTHER EUROPEAN COUNTRIES

5.1 The Czech Republic

5.1.1 Introduction

The Czech Republic is a major producer and exporter of electricity. Generation in 2012 was 87.57 billion kW-hour [1]. It is noted in [2] that in the Czech Republic a distinction is made between lignite and brown coal, whereas almost everywhere else the terms are used synonymously and interchangeably. Even so they are lumped together for the purposes of the analysis in [1]. The distinction is on the basis of estimated geological age, not of properties such as calorific value and moisture content. The section will also take in the Slovak Republic, where such a distinction also applies as it does in parts of the former Yugoslavia (see section 5.8.1).

5.1.2 Lignite deposits

Seventeen lignites from the Czech Republic are discussed in a fairly recent addition to the literature [3]. These are classified into sapropelic coal, liptobiolith coal, xylite-rich coal and matrix coal. 'Sapropelic coal' and 'liptobiolith coal' are distinguished from the 'humic coals' which constitute the peat to anthracite series [4]. Liptobiolites are derived from material such as pollen, spores and resins. In [3] as in other work in the area sapropelic coal and liptobiolith coal are classified as lithotypes along with xylitic coal and matrix coal, the latter being a widely used term denoting amorphous material also called groundmass. Sapropelic coal and liptobiolith coal both contain mineral matter. In fact kerogen, the organic component of shale which can be retorted to make oil products, is sapropelic. The 17 coals studied in [3] were taken from five deposits, and the approach taken in this section will be a description of power generation with coal from the respective deposits. Sapropelic coals often occur as layers within humic coals (see section 5.2.3). There is sapropelic coal at the Austrian deposit discussed in section 4.4 and other sapropelic coals are identified in later parts of the book. Sapropels, the precursor to sapropelic coals, are discussed in Chapter 21 on account of their analogy to peat.

5.1.3 Prunéřov power stations

Note the plural in the heading. Prunéřov comprises two power plants significantly spaced in age: Prunéřov I which commenced operations in 1967–68 and Prunéřov II which commenced operations in 1981–82. Prunéřov is now operated by ČEZ (headquarters in Prague). It is fuelled by brown coal from the Nástup Tušimice mines which are part of the North Bohemian Basin [5,6]. Prunéřov I has an output of electricity of 660 MW between three equivalent units. Prunéřov II has 1050 MW, between five equivalent units. The Nástup Tušimice mines produce 13.5 million tonnes of lignite annually. Noted in [5] as well as in other sources is the need to homogenise the lignite from this source before milling it to make p.f. There has been recent introduction of supercritical steam at Prunéřov. The enhanced efficiency which supercritical steam provides for obviously reduces greenhouse gas emissions.

At the Tušimice mine brown coal is homogenised. The method works by stacking the coal vertically and removing it for use from horizontal layers. It is not difficult to understand in semi-quantitative terms how this homogenising process works. Vertical drop in stacking will provide some segregation of particles according to their mineral matter content, as mineral matter is more dense than organic 'coal substance'. This means that in a stack so created a horizontal layer will have much less variation in density than the coal before stacking, which is why there is removal of the coal for boiler furnace entry in horizontal slices as noted. The higher the mineral matter content, a measure of the non-combustibles within a coal particle, the lower the calorific value. The effect of inhomogeneity of the coal input would be flame instability. Even where the coal is not sufficiently inhomogeneous to require special measures this can happen, and a common solution is to use briquettes instead of or additionally to unprocessed coal for as long as it takes to stabilise the flame at the p.f. burner. In such a contingency the briquettes are of course milled before admittance to combustion plant.

Each of the Prunéřov power stations has sulphur dioxide release control by reaction of the sulphur dioxide in the flue gas with limestone. This is a classical but by no means obsolete means of sulphur dioxide emission control. This control measure at Prunéřov has been in place since the mid 1990s, a few years after the new regime in the Eastern bloc countries came into being.

5.1.4 Hodonin power station

This is one of the oldest power plants in the Czech Republic, having come into service in the 1950s [7]. Its siting was intentionally close to a lignite deposit, though this fuel was unsatisfactory and sometimes had to be replaced by lignite from Sokolov in northern Bohemia, one of the deposits featuring in [3]. The difference between the two coals was probably in their grindability, a quantity routinely determined for coals of all ranks by means of the Hardgrove index as explained in Chapter 1. This is of course relevant to the milling of coal to obtain p.f. Important to p.f. performance in terms of flame stability is the particle size distribution and a particular distribution, notably the Rosin–Rammler distribution [8], is often aimed for. It might have been that at Hodonin initially the milling conditions gave a good particle size distribution with the coal from Bohemia but less so with that from Moravia. It is noted in [7] that the difficulties were in drying and milling, supporting the hypothesis above that grindability was the origin of the difference.

The company Lignit Hodonin, which had a monopoly in lignite production, went into insolvency and as a result there was no production of lignite in the Czech Republic in 2010 [9]. At the present-day Hodonin power station there are, in addition to the p.f. boilers, two boilers using fluidised beds. Fluidised beds using coal of any rank require particle sizes of the order of 1 mm, an order of magnitude higher than the particles in p.f. The fluidised beds at Hodonin have used lignite–biomass co-firing, and now one of the fluidised beds uses biomass alone. This is consistent with the view that co-existing coal and biomass combustion has an important role in carbon dioxide reductions and that this stimulates rather than restricts developments in coal utilisation. Moravian lignite also features in section 15.6.

5.1.5 Opatovice power station

This uses fuel from the central part of the North Bohemian Basin (NBB) [10]. Combined heat and power (CHP) is practised at this power station, which in 2008 (the most recent year for which the information is available) produced 2241 GW-hour or, equivalently, produced at:

$$\mathbf{2241 \times 10^9\ J\ s^{-1}\ hour/(24 \times 365)\ hour = 255\ MW}$$

and the nameplate capacity is about 40% higher than this [11,12]. There are two new desulphurisation units at Opatovice [12], each of which uses lime.

5.1.6 Other major lignite-fired power stations in the Czech Republic

These include the Melnik power station [13], which is the closest in distance to the national capital Prague and uses fuel from the NBB. Built in parts coming into service at quite widely separated times, it has a capacity of 1072 MW. There is also the Komořany power station, capacity 239 MW, which is situated at Most in the north-west part of the Czech Republic close to the border with Germany and uses fuel from Tušimice [14]. The Nováky Power Plant is the only such utility in the Slovak Republic to use brown coal, and it is local. It is noted in [15] that the Slovak Republic has a billion tonnes of brown coal. Even so > 20% of that used is imported from the Czech Republic. Analysis figures of Slovakian brown coals collectively are given in [15] and these include water contents in the range 24–36%. Brown coals originating in Slovakia are up to 2.5% sulphur. This is reflected in the fact that in the 1990s new desulphurisation plant was installed at the Nováky Power Plant, bringing the standards in terms of sulphur dioxide release to those in western Europe.

5.2 Poland

5.2.1 Introduction

The same approach will be taken as for the Czech Republic: background on the nature of Polish lignites drawing on suitable recent literature followed by linkage of the information on the lignites to their use in power generation at power plants selected for discussion.

5.2.2 Polish lignites

In [16] the order of reactivity for maceral groups:

vitrinite/huminite > liptinite > inertinite

is given as well-known background, and the identity of vitrinite with huminite previously noted in this book applies. The lignites studied were from deposits in the 'Polish Lowland' and were found to be typically 89% in huminite.

5.2.3 Bełchatów power station

Having commenced operations in the 1980s, Bełchatów now delivers 5 GW of electricity and supplies a fifth of Poland's electricity. Situated near Lodz, it has established itself as the largest non-nuclear power station in Europe. It uses p.f. obtained from lignite from the Polish Lowlands at a mine known as the Bełchatów mine. The fly ash from burning the lignite as p.f. is micron or sub-micron in particle size. Fly ash can sometimes be used as an ingredient for manufacture of cement, concrete and bricks. It is reported in [17,18] that fly ash from Bełchatów holds promise for such use. This is probably due to the calcium content. Fly ash from black coal tends to be more suitable for cement production than that from brown coal [19]. It is worth noting that in bids for building contracts for the US Federal Government use of materials containing fly ash is seen as a plus.

Bełchatów lignite is a major reserve having been the subject of investigations varied in nature and reported in the peer-reviewed literature. In [20], which is concerned with methane sorption by Polish lignites, we are told that its maceral analysis is dominated by huminite (consistently with what is reported in [16]) and that there is significant xylite. The need – common to usage of all lignites – to effect drying is the subject of [21] in which superheated steam at 110–170°C is used on a test scale to dry Bełchatów lignite in spherical particles of 5 mm and of 10 mm, representing of course particles before pulverisation. It was noted that formation of cracks in the particles altered their surface areas, and that this was a factor in drying performance. This theme continues in [22] in which, for the drying of a Bełchatów lignite, equilibrium between applied steam and coal moisture content is examined. It is clear from these few arbitrarily chosen examples that advances in lignite utilisation are being applied to the coal from Bełchatów, which has an important role in Poland's future energy supply. The term sapropelic was introduced earlier in the chapter in the description of Czech lignite. The deposit at Bełchatów contains some sapropelic lignite as layers within humic coal, as well as xylite [23].

Returning to the power station at Bełchatów, it emits 30 million tonnes per year of carbon dioxide [24]. Mitigation measures are in hand, and they include an advanced process involving reaction of the carbon dioxide with amine, intrinsically a standard approach. A photograph of the Bełchatów power station can be found in [24].

5.2.4 Pątnów power station

Expanded in stages since its initial use in 1967, this power station now has a capacity of almost 1.7 GW.

Pątnów uses lignite from the Konin mine in the Zittau basin in central Poland [26].[8] Coal from Konin as mined has a calorific value of 9 MJ kg^{-1} and is high in sulphur.

8 The Zittau basin extends across more than one country, and enters North Bohemia.

As with Germany and the Czech Republic, by drawing on hopefully judiciously chosen research literature some facts about Konin lignite will be gleaned. In its lithotype analysis [27], bands in a vertical section of the coal deposit of the order of 20 cm in width are assigned names as follows. There is xylitic lignite, a term having previously featured in this text. There is detritic lignite with an internally highly consolidated ('massive') structure though not free from fractures, also detro-xylitic lignite and xylo-detritic lignite. In fact the coal studied in [27] was 'in the vicinity of Konin' and the same is true of the work in [28] which treats Konin as a 'basin' in the 'midlands of Poland' (see also section 5.2.6). It is to the east of the Polish Lowlands previously referred to. Additionally to analysis on the basis of detritic lignite and xylitic lignite and so on, [28] gives huminite reflectances for 18 lignites from the 'Konin basin'. These are in the range 0.16–0.27. The fly ash from Pątnów has been evaluated for another fairly common application [29]: blending with soil to the enrichment of elements including phosphorus and calcium. Sometimes this process has the further beneficial effect of raising the porosity of the soil to which the fly ash is added.

5.2.5 Turów power station

This is in south-west Poland and has a capacity of 2 GW. It uses as fuel lignite from the Turoszów basin, which is approaching 15% in ash. The set-up at Turów is what is known as a mine mouth power station, a self-explanatory term, and the present name-plate capacity is 1698 MW [30].

Turoszów features alongside Konin in reference [28], where it is recorded that 19 lignite samples from Turoszów range in huminite reflectance from 0.18 to 0.35%. A paper concerned with pyrolysis of a group of low-rank coals [31] gives as background the calorific value of a Turoszów lignite on a dry, ash-free basis as 18.6 MJ kg^{-1}. The same source gives the volatile matter content as 48.6%. The ash content is given as 10.2%, lower than for some coals from Turoszów. On the pollution front, it is stated in [32] that at a time about ten years earlier emissions of lead from Turoszów had harmed tree life.

There is at Turów concurrent use of p.f. and several fluidised beds [33,34], and the steam generated by the fluidised bed is in the supercritical state. Turów was one of the earliest power plants to use supercritical steam raised by fluidised bed combustion. The power station began operations in 1962, so renewal as well as expansion are taking place. A new 430–450 MW unit is to be built [35] and a unit of approximately equivalent rating is to be withdrawn from service. The new unit will of course receive lignite from Turoszów.

5.2.6 Adamów power station

This, like Turow, is getting long in the tooth having been generating since 1966. It draws lignite from the Adamów mine, adjacent to Konin; they are sometimes referred to as the Konin-Adamów basin [36]. Details for Adamów lignite amongst others from Poland are given in [37]. Such details include the fact that it is high in huminite and also that it is low in ash. In [20], previously cited in relation to Bełchatów lignite, it is stated that the petrographic analysis of Adamów lignite is 75% and that the huminite reflectance is 0.27–0.28%.

The Adamów power station does have a limited future: it is expected to cease operations in 2024 [38][9], which is probably why there has been no recent expansion.

5.2.7 Proposed Legnica power station

There are plans for a new power station at Legnica in Lower Silesia [39] which, if it is completed, will draw lignite from a deposit of the same name. Irrespective of whether or not the Legnica power station comes to fruition, the Legnica lignite mine is seen as an important supplier of fuel for the future power industry in Poland [40]. In terms of lithotype analysis, Legnica lignite is detritic with a small amount of xylite (see section 5.2.3) and it is high in huminite [41]. That the Legnica power station proposal will go ahead is not certain. Even more speculative is power generation at Gubin-Mosty in western Poland, where there is a deposit of quite sufficient size to supply a major power station.

5.2.8 Concluding remarks

Heavy dependence on lignite in post-communist Poland is clear. It is increased by a very striking point regarding Poland's energy mix: there are no nuclear power plants in Poland. By contrast the Czech Republic – also formerly 'Eastern Europe' – has two nuclear power stations, one with four reactors and the other with two. These have a combined capacity of 3.8 GW. Hungary (see section 5.5) receives 46% of its electricity from its nuclear power station at Paks. So whither Poland in the matter of lignite production utilisation?

5.3 Greece

5.3.1 Introduction

We are informed in [42] that Greece 'boasts' 4.7 billion tonnes of lignite and that in 2012 electricity was generated from it at a rate of 11.2 GW, this being 68% of the total rate of generation in that year. Below, selected major power stations will be discussed in turn as for Germany, the Czech Republic and Poland above.

5.3.2 Ptolemaida power station

This was the first such facility in Greece to use lignite as fuel, and will accordingly be described below. It entered service in 1959 [43] which represents a late start; there were lignite-powered power stations in other countries long before then. The power station produces at 495 MW and a further unit which will more than double that is expected to start producing in 2019. The Florina power station, close to Ptolemaida, produces

9 'Poland remains committed to lignite power generation', Energy Post 17 November 2014. Accessible at: http://www.energypost.eu/wishing-away-lignite-eu-climate-policy-ignores-elephant-room/

330 MW. The mine of the same name, though a.k.a. the Ptolemaida-Florina Mine, produces 49 million tonnes per year of lignite. Lignite from the deposit is of low calorific value, 5–6 MJ kg^{-1} when *in situ* at the mine [44]. This is a direct consequence of the unusually high ash content, reported in [45] as being 38.4% dry basis. Use in power generation of coal from this basis produces annually 8 million tonnes of fly ash [44].

The new unit at Ptolemaida will occupy space at a now depleted part of the lignite deposit. Construction of the new unit at Ptolemaida will be by Hitachi [46], who have constructed new capacity for a number of lignite-fired power stations including two having featured previously in this text: Lippendorf and Boxberg. Opposition to the new unit was inevitable and in a protest by World Wildlife Fund [47] the point is made that the expected efficiency of 40.6% in the new unit at Ptolemaida (obtained of course by use of supercritical steam) is below the efficiencies which are realised at some Vattenfall and RWE generating facilities using lignite. This argument merits close consideration, but that is not true of such utterances as that in [48] which describes the extension at Ptolemaida as being 'anachronistic', following that with a header 'Lignite has serious impacts'. This disregards the obvious point that standards of emission control can be high or low with any fuel and that the term 'clean coal' has more to do with combustion practices and emission control than with the coal itself, a point reiterated in section 16.4.

5.3.3 Agios Dimitrios power station

Having entered service in 1983 at a capacity of 600 MW, this facility now has five 300 MW turbines, the nameplate capacity of 1.5 GW making it a 'big player' amongst the power plants in the Mediterranean countries. It utilises 65–70 thousand tonnes per day of South Field lignite [49]. As in section 4.2.6 which dealt with lignites from Lusatia, mine dumps will feature in the discussion of South Field lignite.

Over the period 10 May to 15 July 2004, there was movement of a pile of waste at South Field [50]. Internal destabilisation is believed to have involved movement of the order of 40 Mm^3 of matter in the pile. A quantity equal to about a fifteenth of this exited the dump zone. Factors in the incident were the high water content of the substances within the pile and exacerbation of water effects by the existence of a spring at the base of the pile [50].

As mined, South Field lignite has a calorific value as low as about 5 MJ kg^{-1}, again because of the ash content which ranges from 28 to 44% [51]. An interesting point arises in relation to sulphur in lignites at South Field. The sulphur content is about 0.4%. No sulphur dioxide is emitted because there is sufficient calcium in the coal to trap the sulphur as calcium sulphate [52]. This same effect has been reported for Australian brown coals [53].

Recent promotional literature on Agios Dimitrios [54] says: 'Lignite is a somewhat inefficient fuel due to its low heat value.' That of course applies to Greek lignites in particular because of their high ash contents as noted. The statement continues: 'Still, having [lignite] mines of our own nearby has an advantageous effect on the production costs.'

5.3.4 Megalopolis power station

This draws on a lignite mine of the same name as well as on two other mines. Its expected production for 2020 [55] is 6 million MW-hours equalling, in round-the-clock operation, a generation rate of:

$$6 \times 10^6 \text{ MW-hour}/(365 \times 24) \text{ hour} = 685 \text{ MW}$$

Megalopolis has five mills for pulverising, each capable of delivering 180 tonnes per hour. Electricity for this power station services a number of the Greek islands.

A good deal has been written about radioactivity at the deposit from which the Megalopolis power station sources its lignite, so it is appropriate in this section to discuss this issue in general terms with Megalopolis as the background. There is often measurable radioactivity in lignites, originating with the mineral content. In Victoria, Australia, a device was once developed which guided the operators of the dredger at a brown coal open cut. The principle of operation was that a radioactive meter fitted to the dredger would give a response dependent on the mineral matter of the coal being accessed, which will vary across the open cut [56]. The operators were thereby alerted if the dredger had entered an area particularly high in mineral matter, and could re-position it so as to avoid coal high in mineral matter and therefore productive of ash on burning.

The focus of investigations of these effects at Megalopolis has been radioactivity in the soil close to the power station itself and at the scenes of fly ash disposal [57]. Measurements were taken at six sites none more than 5 km from the boundary of the power station, which comprises five units all using lignite as fuel. A little necessary background on units and definitions in radioactivity is given in the box.

- The term 'Roentgen equivalent man' (rem) is 'the dosage rads that will cause the same amount of biological injury as one rad of X rays or gamma rays' [58].
- One rad is the dose which causes 0.01 joule of energy from the radiation to be absorbed per kilogram of the (usually human) receiver.
- One Sievert (Sv) is 1 J kg^{-1} of received energy, resulting from a dose of 100 rad.
- With a radionuclide, 1 Becquerel is one disintegration per second. Signifying one event per second it is equivalent to unit frequency viz. 1 s^{-1}.

In the study at Megalopolis, the six sites ranged from 180 to 500 nSv $hour^{-1}$ of gamma radiation. Measurements were taken 1 m above the ground. There were also measurements of three particular radionuclides from soil taken from each of the sites: ^{226}Ra, ^{232}Th and ^{40}K. Across the six sites ^{226}Ra ranged from 80 to 593 Bq kg^{-1}, ^{232}Th from 32 to 44 Bq kg^{-1} and ^{40}K from 300 to 524 Bq kg^{-1}. Note that with the radionuclides the measurements are per kilogram of releaser not per kilogram of receiver as with the gamma ray measurements.

The particular study being examined [57] concluded that scenes of fly ash disposal did not exceed in radioactive content adjacent areas where there had been no such disposal. There are on the internet innumerable sources of information on gamma radiation emission levels and radionuclide disintegration levels from particular substances and at particular locations, and these the interested reader can readily access. He or she can draw on the contents of this section for an introduction to the formalism of radioactivity and for a 'feel' for the dosage and emission rates which would apply to fly ash.

5.3.5 Amyntaio power station

Known locally as the Filotas Steam Electric Station, this produces at 600 MW from two equivalent units. Its emission factor for carbon dioxide for 2011 is given as 1349 kg per MW-hour of electricity [59], and this is compared with the values for some other Greek lignite-fired power stations including Ptolemaida (1577 in the same units) and Florina (1210 in the same units). By way of comparison, the factor across coals of all ranks in the USA is 1020 kg per MW-hour [60].

Simple calculations apropos of the carbon dioxide emissions at Amyntaio are presented in the box.

600 MW round-the-clock is annually:

[600 MW × (365 × 24) hour] MW-hour, productive of:

[600 MW × (365 × 24) hour] × 1349 × 10^{-3} tonne of carbon dioxide

= 7 090 344 tonnes of carbon dioxide according to the emission factor of 1349 kg $(\text{MW-hour})^{-1}$ given in [59].

In [61] the total carbon dioxide release for the two 300 MW (= 600 MW) units at Amyntaio is given as <u>5 124 545 tonnes</u>, a figure 28% lower, simply indicating that the units are not working at full capacity the whole of the time.

Data from the two sources [59,61] therefore hang together.

That there are serious environmental difficulties in the Municipality of Amyntaio, home to the 'Filotas Steam Electric Station', is clear from [62], in which it is pointed out that there was in the area no investment in renewables until 2009.

5.3.6 Kardia power station

This produces at 1250 MW, using in p.f. form coal from the Kardia field of calorific value 5.5 MJ kg^{-1} and 31.5% ash content. Superheated steam is used. In a study from 2007, twenty-six samples of lignite from the Kardia field were examined [63]. In the air-dried state, the samples ranged from 9 to 25% moisture. On a dry basis, ash ranged from 13 to 29%. Carbon dioxide from devolatilisation at 550°C ranged from 1 to 16% dry basis. Dominant constituents of the ash included SiO_2 and Al_2O_3. Arsenic was surprisingly prevalent as a trace element, up to 127 μg per g coal. Strontium was present in even larger amounts. In view of the high ash content, that slagging and fouling are a difficulty is no surprise and the deleterious effects of these at the Kardia power station are reported in [64]. There are electrostatic precipitators there which were upgraded in the early 2000s [65]. Kardia and Agios Dimitrios have come under criticism (e.g. [66]) on account of their emissions.

Returning to the matter of devolatilised carbon dioxide, this arises from the fact that the oxygen content of a lignite represents parts already oxidised and is conceptually distinct from that arising from combustion. This has to be accounted for in a mass balance. In section 1.4 it is mentioned that two particular drying procedures for low-rank coal also achieve decarboxylation. This is of course beneficial in carbon dioxide emission terms.

When a lignite from the Amynteon-Ptolemaida deposit was studied for petrographic properties [67] it was concluded that within the inertinite group of macerals there was both autochthonous and allochthonous material. The 10% of inertodetrinite was judged to be allochthonous whilst the remainder of the inertinite macerals were judged to be autochthonous.

The gelification index (GI – see section 2.2) for coal from this source was also reported. It was defined as:

GI = (ulminite + humocollinite + densinite)/(textinite + attrinite + inertinite)

Whilst the huminite macerals are lumped together in the denominator, all of the other quantities are for particular macerals, and the calculation was performed for 20 samples from the deposit. As an example, one sample was inertinite 0.5%, attrinite zero, textinite 6.8%, ulminite 60.6%, densinite 17.4% and humocollinite 1.9%. Substitution gives a GI value of 10.9. By contrast another sample was inertinite 57.5%, attrinite zero, textinite 0.3%, ulminite 15.0%, densinite 18.4% and humocollinite 1.9%, giving a GI value of 0.6. In the first the degree of gelification is high and in the second low. The cut-off between highly and less highly gelified coals was set at GI = 2, and all of the samples except two were highly gelified on this criterion.

5.4 Bulgaria

5.4.1 Introduction

The leading lignite mining operator in Bulgaria is Mini Maritza Iztok Ead [68], the country's largest employer. It supplies the power stations discussed below. Some of the properties of Maritza lignite are summarised in [69], in which we are told that the ash content on a dry basis is 14.5%. There is a high proportion of sulphur – organic and as pyrite – which together account for over 3.5% of the coal weight. The calorific value on a dry, ash-free basis is 25.5 MJ kg^{-1}. On the petrographic front, the following macerals have been separated from lignites from Maritza [70]: resinite, humoresinite, textinite, ulminite, semifusinite and fusinite.[10]

10 A reader might wish to consult reference [26] in Chapter 2, which was cited as a reference for maceral nomenclature.

5.4.2 The Maritza complex

This comprises three lignite-fired power stations: Maritza East 1 ME-1, Maritza East 2 ME-2, and Maritza East 3 ME-3. They all use lignite from the Maritza East mines [71]. They are mine mouth plants.

Power generation at the complex began in the early 1960s. There was once a lignite drying plant, but this went out of service. Instead, raw lignite is combined with briquettes, which are milled together to give p.f. of enhanced calorific value. Maritza East 1 ME-1, 250 km south-east of Sofia [68], currently generates at 670 MW [72]. It is being evaluated for expansion in environmental terms, reflecting Bulgaria's ongoing commitment to lignite.

Maritza East 2 ME-2 has evolved to its current performance and level of capitalisation over a period of a little over 50 years [71]. It has eight generating units with a combined production of 1465 MW. As at other power plants featuring in the book, gypsum is a saleable product from the desulphurisation of the flue gases. Maritza East 3 ME-3 has been in service for 35 years [73]. Its expected electricity production for 2020 [73] is 6 049 210 MW-hour which signifies power production round-the-clock of 690 MW. In section 5.3.5 the carbon dioxide emission factor for the Amyntaio power station in Greece was examined and a value for power stations in the USA was also given. Reference [73] gives a predicted 6 163 230 000 kg of carbon dioxide from Maritza East 3 ME-3 in 2020. This gives a carbon dioxide emission factor of 1018 kg per MW-hour of electricity. This compares favourably with the value of 1349 kg per MW-hour of electricity for the Amyntaio power station and just about equals the value of 1020 kg per MW-hour for power plants using coal of all ranks across the USA (see section 5.3.5).

5.4.3 The Bobov Dol coalfield and power generation there from

This reserve in south-west Bulgaria consists of 100 million tonnes [68]. There are differences in reporting of its rank. For example, it is described in [68] as a brown coal and in [74] as a sub-bituminous coal.

Power generation at Bobov Dol commenced in 1974 [75]. Its projected supply for 2020 is 1 999 300 MW-hour (= 228 MW of power) and its expected carbon dioxide emission factor for the same year is 1187 kg per MW-hour of electricity.

5.5 Hungary

5.5.1 Brown coal deposits

In Hungary brown coal far exceeds higher ranks of coal in abundance [76]. The three scenes of brown coal mining in Hungary are Visonta, Bükkábrány and Márkushegy: the first two are open cut and the third underground. Márkushegy provides not more than 10% of the total electricity. Visonta brown coal is 25% in ash, with silica the dominant constituent of the ash [77]. Its calorific value *in situ* at the mine is 6.6 MJ kg^{-1}, and sulphur is 0.86% [78]. Márkushegy is another example of uncertainty over rank: classified in [76] and in other sources (e.g. [79]) as brown coal or as lignite, it is seen by the authors of [80] as being sub-bituminous. A reserve can display sufficient variations in properties for such uncertainties to arise, a point which emerges several times in this book including section 10.3 where, with an Australian low-rank coal as an example, the ASTM[11] classification of coal ranks is applied to a reserve showing variation.

5.5.2 Matra power plant

This is in Visonta and uses lignite from the mine there and from Bükkábrány. An output of 876 MW is realised by five non-equivalent units using lignite, and 66 MW is available as required by way of top-up from a turbine using natural gas. It was announced in March 2015 that the major international concern Alstom (headquarters near Paris) had been awarded a contract to renovate several of the steam turbines at Matra [81]. Here again commitment to continued lignite usage is evident. Indeed, extended life for Matra is being ensured in a number of ways [82]. These include new desulphurisation units and modified burners to control NO_x and SO_2 emissions. There are also plans for ash disposal in the following way. It will be mixed with an equivalent weight of water to form a pumpable paste, and taken that way to a suitable disposal site. Hungarian lignites feature in two further parts of this book: section 11.2.1 (briquetting) and section 15.3.2 (supercritical extraction).

5.6 Turkey

There is in the literature an abundance of material on Turkish lignites. Across a group of 24 lignites from Turkey examined in [83] ash contents ranged from 7.6 to 43.3% and calorific values on a dry basis from 15.63 to 30.37 MJ kg^{-1}. It is noted in [84] that only 6% of the known lignite reserves of Turkey – >13 million tonnes – have calorific values exceeding about 12 MJ kg^{-1}.

The planned Soma Kolin power station [85] is being built close to Ankara, and will comprise two 225 MW generating units fired by lignite from Denis (a.k.a. Soma Denis), Evciler, Kozluören and Türkpiyale. The lignite at Soma Kolin will be burnt not as p.f. but in fluidised beds, heat from which will raise steam for turbine passage, and there will be flue gas desulphurisation. There is both a hopper and an electrostatic precipitator for ash control. There is a 457 MW lignite-fired power station in Sivas, central Turkey, which draws on the Kangal lignite deposit [86]. The lignite at Kangal is hypautochthonous [87], there having been movement within the bed of deposited plant debris.

There is a return to the burning of lignites from Turkey in section 20.2, where their co-firing with biomass is considered.

11 American Society for Testing and Materials. See also section 24.1.

5.7 Romania

Romania mines black coal and lignite, and all of it is used in power generation [88]. Selected lignite power stations are described in Table 5.1.

Table 5.1 Lignite-fired power stations in Romania.

POWER STATION	DETAILS
Turceni	2310 MW of electricity, the largest power plant in Romania [89]
Rovinari	Four 330 MW units [91]
Mintia-Deva	1285 MW [92]
Craiova	300 MW [93]

Turceni (row 1) is a major polluter, and applications for financial aid to modernise and upgrade have run into difficulties [90]. Alstom (see section 5.5.2) are to provide flue gas desulphurisation for Rovinari (next row). Mintia-Deva began producing in 1969, and has been expanded several times. Over that time it has produced 194 TW-hour of electricity, giving a mean production over an approximate 45-year period of:

$$[194 \times 10^{12} \text{ W-hour}/(45 \times 355 \times 24) \text{ hour}] \times 10^{-6} \text{ MW} = 500 \text{ MW}$$

or just under 40% of its current nameplate capacity. Alstom are also involved at Craiova (following row) where, as at Rovinari, they will install flue gas desulphurisation.

5.8 The former Yugoslavia

5.8.1 Introduction

In Yugoslavia before re-organisation the southern regions were scenes of lignite production [94], and lignite is more abundant within the borders of the former Yugoslavia than bituminous coal. The most important centre of activity was the Tito mines complex in Tuzla, now in Bosnia and Herzegovina. It was noted in section 5.1.1 that within the former Yugoslavia a distinction is made between lignite and brown coal, as it is in the former Czechoslovakia.

5.8.2 Bosnia and Herzegovina

There is large-scale use in Bosnia and Herzegovina (BH) of lignite in electricity generation [95]. The power generator Elektroprivreda BiH (EPBiH) owns power plants at Tuzla and at Kakanj of capacity 715 and 460 MW, respectively, and each uses lignite fuel. Elektroprivreda RS (EPRS) owns Ugljevik (300 MW) and Gacko (300 MW) power plants, also lignite fired. The power plant at Tuzla entered service in 1959. New units were added and some in turn decommissioned [96]. Tuzla obtains its lignite from a number of sources, and it is co-fired with 25% of black coal [97]. Kakanj coal is extremely high in ash at 41%, though only moderate in sulphur (0.26%) [98]. Some of the ash is diverted to cement manufacture [99] as at Bełchatów, above. Coal from Ugljevik is as high as 4–6% in sulphur [95] whilst that from Gacko is around 1% [100]. Lignite production at Dobrnja in BH, and the grave accident there in 1990, are discussed in Chapter 17.

5.8.3 Slovenia

The most important lignite mine in Slovenia is that in the Šaleška Valley, a.k.a. the Velenje basin. This supplies the Šoštanj power plant. It came into operation over 50 years ago, and the units initially present are closed down. A new unit recently came into operation which uses supercritical steam [101].

Until 2013 an underground lignite mine at Trbovlje, Slovenia was producing [102] and supplied the power station of the same name. Entering service in 1966, this power station has a capacity of 250 MW from steam turbines plus an extra 64 MW from two gas turbines [103]. The term sapropelic coal was introduced in section 5.1.2 and, in section 5.2.3, applied to lignite from Bełchatów. Lignite from Trbovlje is also sapropelic [104].

5.8.4 Serbia

Table 5.2 gives details of some lignite-fired power stations in Serbia.

Table 5.2 Power stations in Serbia.

LOCATION AND REFERENCE	DETAILS
Kostolac [105]	A new unit installed which, using supercritical steam, will achieve an efficiency of 40.8% and produce 600 MW of electricity
Kolubara [106]	245 MW from lignite. Plans for Kolubara B, a 750 MW facility, unfulfilled [105]
Nikola Tesla [108]	The largest power plant in Serbia. Six units. Lignite from Kolubara
Morava [109]	Entered service in 1969. Capacity 125 MW. Lignite from Kolubara and from Rembas (see main text)

Lignite for the Kostolac power station is from the basin of the same name, and petrographic details of this are given in [107]. Across a range of 13 lignite samples from the deposit the huminite content, which in [108] is resolved into individual macerals, varies from 60.7 to 74.0%. The Hardgrove index is remarkably consistent at 41–45. The same details are given for Kolubara coal. They are, for a range of 11 samples, total huminite 57.2–85.7% and Hardgrove index 35.8–56.9.

The annual production at Nikola Tesla is 8 billion kW-hours [109], giving a production rate of:

$$8 \times 10^{12}\ \text{W hour}/(365 \times 24)\ \text{hour} = 915\ \text{MW}$$

which can be compared with the installed capacity of 1650 MW given in [108]. The Morava power plant (next row) has been producing steadily for 45+ years though at a low rate. There are proposals for a lignite-fired power station at Despotovac in Serbia, which will use lignite from the Rembas mine [110] some of which will go to Morava.

5.8.5 Kosovo

Kosova (note the 'a') A and B power stations, both in Obilić, use as fuel lignite from the Kosovo basin. Kosova A and B have a combined capacity of 1478 MW [111]. Lignite is by far the biggest contributor to electricity generation in Kosovo. Lignites are not all of the same degree of coalification or maturity, and vitrinite or huminite reflectance is a measure of this. By geochemical analysis lignites from the Kosovo basin were deduced to be less mature than many [112].

5.8.6 Montenegro

Pljevlja is the only lignite-fired power station in Montenegro, providing about 30% of the country's electricity needs [113]. Its capacity is 210 MW, enabling it to produce in a year:

$$\mathbf{210 \times 365 \times 24\ MW\text{-}hour = 1.8\ TW\text{-}hour}$$

Its actual annual output is given as 1.1 TW-hour. There have been proposals for expansion [114]. Compliance with EU limits for emissions from power plants is one of many issues to be faced by Montenegro before her admittance to the EU becomes a reality.

5.8.7 Macedonia (Republic of)

Lignite-fired power stations at Bitola and Oslomej contribute 800 MW to the electricity generation capacity of the Republic of Macedonia [115]. The lignite is obtained from the mine at Suvodol. Suvodol lignite and Florina lignite (above) have been compared for performance in power generation, especially with regard to emissions [116].

5.8.8 Croatia

Croatia does not currently produce lignite, though the part of Yugoslavia which is now Croatia did up to *c.*1975, at mines at Dobra and Istria [117]. It was used at the Plomin power station, and supply of lignite to there by import from Kolubara has been discussed without yet being a reality.

5.9 Italy

A notable lignite reserve is that at Sulcis, Sardinia [118]. Under the Mussolini regime, when Sulcis was producing > 1 million tonnes annually, a town nearby was renamed Carbonia. Proximate analysis figures for Sulcis coal are given in [118] as volatile matter 35.6%, fixed carbon 40.1%, moisture 8.2% and ash 16%. The ash then is moderately low. The petrographic analysis, using figures in [119], is vitrinite 82.0%, liptinite 12.3%

and inertinite 5.6%. Its vitrinite reflectance is given in [120] as 0.48%. Sulcis is another lignite sometimes classified as sub-bituminous like those from Bobov Dol (see section 5.4.3) and that from Márkushegy (see section 5.5.1) amongst others. In [119] it is concluded, on the basis of infra-red analysis and the occurrence of bands signifying humic acids, that Sulcis coal is a lignite.

There is also lignite at Baccinello in Tuscany, central Italy [121], a comment on which is made in section 21.2. Lignite from the Tiberino basin, also in central Italy, has been characterised as being autochthonous [122].

5.10 Other European countries

Spain no longer produces lignite, although it does feature later in the book for example in section 15.3.2 where supercritical extraction is the topic. In Catalonia there is a thin deposit of lignite known to be allochthonous [123]. It has plant debris identifiable with roots, often a characteristic of an allochthonous deposit. Detailed information on Mequinenza lignite from north-east Spain was published a little over 20 years ago [124]. Its volatile matter on a dry basis is 43.3%, its ash 21.6%. Its huminite reflectance is 0.35%. It is hypautochthonous to autochthonous (see section 2.6).

Portugal features later in the book when briquettes are discussed. Northern Ireland has over a billion tonnes of lignite, and its use to make electricity has long been under discussion [125]. Lignite recently discovered in France [126] awaits characterisation. At Aix Island, off the French west coast, the existence of lignite was first noted in 1817. Gardanne lignite from southern France has been reported as having a vitrinite reflectance of 0.48% [127] and has been described as a 'high rank lignite' [128]. At Arjusanx in south-west France there is a lignite mine where possible utilisation has been discussed [129]. The lignite there is less mature than that at Gardanne.

The Limburg coalfield in the Netherlands once produced lignite amongst other ranks of coal [130]. Lignite from Limburg is now depleted. Such lignite as is known to exist in the British Isles is discussed in Chapter 23, and the discourse now moves to North America.

5.11 References

[1] http://www.indexmundi.com/czech_republic/electricity_production.html

[2] *Czech Republic: Inventory of Estimated Budgetary Support and Tax Expenditures for Fossil Fuels*, Organisation for Economic Co-operation and Development, Paris (2010)

[3] Havelcová M., Sykorová I., Trejtnarová H., Šulc A. 'Identification of organic matter in lignite samples from basins in the Czech Republic: geochemical and petrographic properties in relation to lithotype', *Fuel* 99: 129–142 (2012)

[4] Dexin H. *Coal, Oil Shale, Natural Bitumen, Heavy Oil and Peat. Volume 1: Coal Geology*, EOLSS (2010)

[5] http://www.cez.cz/en/power-plants-and-environment/coal-fired-power-plants/cr/prunerov.html

[6] http://www.sdas.cz/posts/the-nastup-tusimice-mines--doly-nastup-tusimice.aspx

[7] http://www.cez.cz/en/power-plants-and-environment/coal-fired-power-plants/cr/hodonin.html

[8] Jones J.C. *Combustion Science: Principles and Practice*, Millennium Books, Sydney (1993)

[9] http://www.euracoal.com/pages/layout1sp.php?idpage=70

[10] http://www.euracoal.com/pages/layout1sp.php?idpage=70

[11] http://coal.steelguru.com/other_asia/337/eph_claims_czech_coal_supply_contract_termination_unjust

[12] http://www.tractebel-engineering-gdfsuez.com/reference/opatovice-coal-fired-power-plant-desulphurization-unit/

[13] http://www.cez.cz/en/power-plants-and-environment/coal-fired-power-plants/cr/melnik.html

[14] http://www.cez.cz/edee/content/micrositesutf/odpovednost2011/en/environment/vyroba-tezba-a-vystavba.html

[15] Skvarekova E., Kozakova L. 'The issue of brown coal quality on the basis of physicochemical parameters', *The Holistic Approach to Environment* 4: 163–167 (2011)

[16] Bielowicz B. 'Preliminary assessment of the usefulness of the Polish lignite in the gasification process according to its petrographic composition', *64th Annual Meeting of the ICCP*, Beijing (2012)

[17] http://www.cwb-czasopismo.pl/en/?s=11&action=getArticle&aid=481&t=Properties+of+cements+with+calcareous+fly+ash+addition

[18] Hall J.S. 'CSIRO research in ash pelletising', *The Miner* August: 10–11 (1982)

[19] Aalbers T., van der Sloot H.A., Goumans J. (eds) *Environmental Aspects of Construction with Waste Materials*, Elsevier (1994)

[20] Macuda J., Nodzeński A., Wagner M., Zawisza L. 'Sorption of methane on lignite from Polish deposits', *International Journal of Coal Geology* 87: 41–48 (2011)

[21] Komatsu Y., Sciazko A., Zakrzewski M., Kimijima S., Hashimoto A., Kaneko S., Szmyd J.S. 'An experimental investigation on the drying kinetics of a single coarse particle of Belchatow lignite in an atmospheric superheated steam condition', *Fuel Processing Technology* 131: 356–369 (2015)

[22] Kokocinska M., Pakowski Z. 'High pressure desorption equilibrium of lignite obtained by the novel isochoric method', *Fuel* 109: 627–634 (2013)

[23] Worobiec G., Szynkiewicz A. 'Betulaceae leaves in Miocene deposits of the Bełchatów Lignite Mine (Central Poland)', *Review of Palaeobotany and Palynology* 147: 28–59 (2007)

[24] http://www.power-technology.com/projects/belchatow-plant/

[25] http://www.powerengineeringint.com/articles/print/volume-16/issue-7/features/pow-patnow-goes-supercritical.html

[26] http://www.euracoal.org/pages/layout1sp.php?idpage=76

[27] Widera M. 'Lignite cleat studies from the first Middle-Polish (first Lusatian) lignite seam in central Poland', *International Journal of Coal Geology* 131: 227–238 (2014)

[28] Fabiańska M.J., Kurkiewicz S. 'Biomarkers, aromatic hydrocarbons and polar compounds in the Neogene lignites and gangue sediments of the Konin and Turoszów Brown Coal Basins (Poland)', *International Journal of Coal Geology* 107: 24–44 (2013)

[29] http://www.scientificamerican.com/article/coal-ash-in-soil/

[30] http://www.euracoal.org/pages/layout1sp.php?idpage=76

[31] Jones J.M., Kubackia M., Kubica K., Ross A.B., Williams A. 'Devolatilisation characteristics of coal and biomass blends', *Journal of Analytical and Applied Pyrolysis* 74: 502–511 (2005)

[32] Zuna M., Mihaljevi M., Sebek O., Ettler V., Handley M., Navrátil T., Golias V. 'Recent lead deposition trends in the Czech Republic as recorded by peat bogs and tree rings', *Atmospheric Environment* 45: 4950–4958 (2011)

[33] Nowak W., Muskala W., Krzywanski J., Czakiert J. 'The research of CFB boiler operation for oxygen enhanced dried lignite combustion', *10th International Conference on Circulating Fluidized Beds and Fluidization Technology*, Engineering Conferences International (2011)

[34] Johnson F. 'Fluidized bed combustion for clean energy', *12th International Conference on Fluidization – New Horizons in Fluidization*, Engineering Conferences International (2007)

[35] http://www.reuters.com/article/2014/03/18/us-poland-budimex-hitachi-capital-idUSBREA2H1V820140318

[36] http://www.euracoal.org/pages/layout1sp.php?idpage=76

[37] Bielowicz B. 'A new technological classification of low-rank coal on the basis of Polish deposits', *Fuel* 96: 497–510 (2012)

[38] Gawlik L., Grudzinski Z. 'Coal has a future', *Focus on Mining and Power Production* 1(21): 16–19 (2009)

[39] http://www.sourcewatch.org/index.php/Legnica_Power_Station

[40] Ghose A.K., Bose L.K (eds) *Mining in the 21st Century: Quo Vadis? Volume 1*, CRC Press (2003)

[41] Bielowicz B. 'Petrographic composition of Polish lignite and its possible use in a fluidized bed gasification process', *International Journal of Coal Geology* 116: 236–246 (2013)

[42] http://www.euracoal.org/pages/layout1sp.php?idpage=73

[43] http://www.bloomberg.com/news/articles/2012-08-08/greece-s-power-generator-tests-euro-fitness-amid-blackout-threat

[44] Kaldellis J.K., Zafirakis D., Kondili E. 'Contribution of lignite in the Greek electricity generation: review and future prospects', *Fuel* 88: 475–489 (2009)

[45] Koukouzas N., Hamalainen J., Papanikolaou D., Tourunen A., Jantti T. 'Mineralogical and elemental composition of fly ash from pilot scale fluidised bed combustion of lignite, bituminous coal, wood chips and their blends', *Fuel* 86: 2186–2193 (2007)

[46] http://www.hitachi.com/New/cnews/130409a.html

[47] http://www.wwf.gr/ptolemaida5en/

[48] https://www.google.com.au/?gfe_rd=cr&ei=ER4KVYf7DKeN8QeXyYHQAQ&gws_rd=ssl#q=greek+lignite+anachronistic

[49] https://books.google.com.au/books?id=FCNCcCY_pzIC&pg=PA200&lpg=PA200&dq=Agios+Dimitrios+++lignite+sulphur&source=bl&ots=DqGBdE4Xim&sig=0fNOZEoXN-5SopBQVt-2NKM2rly0&hl=en&sa=X&ei=xx7hVP7TD4SymAW-yYG4DQ&ved=0CB0Q6AEwAA#v=onepage&q=Agios%20Dimitrios%20%20%20lignite%20sulphur&f=false

[50] Steiakakis E., Kavouridis K., Monopolis D. 'Large scale failure of the external waste dump at the South Field lignite mine, Northern Greece', *Engineering Geology* 104: 269–279 (2009)

[51] Kavouridis K., Roumpos C., Galetakis M. 'The effect of power plant efficiency, lignite quality and inorganic matter on CO_2 emissions and competitiveness of Greek lignite', *Górnictwo I Geoinżynieria* 31: 355–369 (2007)

[52] *Proceedings of the International Workshop in Geoenvironment and Geotechnics*, Milos Island, Greece (2005)

[53] Jones J.C., Stacy W.O. 'SO_2 and NO_x emissions from the combustion of Victorian brown coals', *Proceedings of the Second Australian Coal Science Conference* pp. 424–430, Australian Institute of Energy (1986)

[54] 'Transformation from the inside', *Automation* 2: 25–27 (2008)

[55] http://enipedia.tudelft.nl/wiki/Megalopolis_Powerplant

[56] Mathew P.J. *The Natural Radioactivity of Brown Coal in the Latrobe Valley and its Application to Exploration and Grade Control in Coal Winning*, CSIRO Division of Mineral Physics, Melbourne (1979)

[57] Karangeloa D.J., Rouni P.K., Petropoulos N.P., Anagnostakis M.J., Hinis E.P., Simopoulos S.E. 'Radioenvironmental survey of the Megalopolis power plants' fly ash deposits', *Bulletin from the National Technological University of Athens*. Accessible at: http://www.google.com.au/url?sa=t&rct=j&q=&esrc=s&source=web&cd=2&ved=0CCIQFjAB&url=http%3A%2F%2Fnuclear.ntua.gr%2Farcas%2Fresearch%2Fradmaps%2Fmeg_deposits.pdf&ei=D03hVMSdE8S6mAWLpoHYAQ&usg=AFQjCNGpPIIdFkZ9KIsrjs-J2SxGw-JOkXLA&bvm=bv.85970519,d.dGY

[58] http://www.nrc.gov/reading-rm/basic-ref/glossary/rem-roentgen-equivalent-man.html

[59] Kaldellis J.K., Kapsali M. 'Evaluation of the long-term environmental performance of Greek lignite-fired power stations', *Renewable and Sustainable Energy Reviews* 31: 472–485 (2014)

[60] http://www.epa.gov/cleanenergy/energy-and-you/affect/air-emissions.html

[61] Vatalisa K.I., Charalampidesa G., Platiasa S. 'CCS Ready innovative technologies in coal-fired power plants as an effective tool for a Greek low carbon energy policy', *Procedia Economics and Finance* 14: 634 – 643 (2014)

[62] 'Rural Web Energy Network for Action', eReNet, March 2013

[63] Adamidou K., Kassoli-Fournaraki A., Filippidis A., Christanis K., Amanatidou E., Tsikritzis L., Patrikaki O. 'Chemical investigation of lignite samples and their ashing products

from Kardia lignite field of Ptolemais, Northern Greece', *Fuel* 86: 2502–2508 (2007)

[64] Panagiotidis I., Vafiadis K., Tourlidakis A., Tomboulides A. 'Study of slagging and fouling mechanisms in a lignite-fired power plant', *Applied Thermal Engineering* 74: 156–164 (2014)

[65] http://www.metka.com/en/projects/projects-of-metka-group/10

[66] http://www.dailymaverick.co.za/article/2012-05-16-wwf-report-save-the-planet-or-start-praying-for-a-new-one/#.VQtRDW8cSUk

[67] Mavridoua E., Antoniadisa P., Khanaqab P., Riegelb W., Gentzis T. 'Paleoenvironmental interpretation of the Amynteon-Ptolemaida lignite deposit in northern Greece based on its petrographic composition', *International Journal of Coal Geology* 56: 253–268 (2003)

[68] http://www.euracoal.org/pages/layout1sp.php?idpage=69

[69] Maes I.I., Mitchell S.C., Yperman J., France D.V., Marinov S.P., Mullens J., Van Poucke L.C. 'Sulfur functionalities and physical characteristics of the Maritza lztok Basin lignite', *Fuel* 75: 1286–1293 (1996)

[70] Siskov G.D., Petrovat R. 'Infrared spectra of coal macerals separated from Bulgarian Iignites', *Fuel* 53: 236–239 (1974)

[71] http://see-industry.com/industrial-statiieng.aspx?br=44&rub=234&id=622

[72] http://www.aecom.com/What+We+Do/Energy/Energy+Planning,+Environmental+and+Economics/_projectsList/AES+Maritza+East+1+Power+Plant

[73] http://enipedia.tudelft.nl/wiki/Maritza_East-3_Powerplant

[74] Vassilev S.V., Vassileva C.G. 'Comparative chemical and mineral characterization of some Bulgarian coals', *Fuel Processing Technology* 55: 55–69 (1998)

[75] http://enipedia.tudelft.nl/wiki/Bobov_Dol_Powerplant

[76] http://www.euracoal.org/pages/layout1sp.php?idpage=74

[77] Barna J. 'Humic substances-clay complexes in Hungarian coals', *Fuel* 62: 380–388 (1983)

[78] Meggyes A. 'NO_x and SO_2 emissions of Hungarian electric power plant boilers', *Perjodica Polytechnica Ser. Mech. Eng.* 35: 177 (1991)

[79] http://www.energiapolitika.eu/index.php/en/a-project-en/background-of-the-project-3

[80] Bechtel A., Hámor-Vidó M., Sachsenhofer R.F., Reischenbacher D., Gratzer R., Püttmann W. 'The middle Eocene Márkushegy subbituminous coal (Hungary): paleoenvironmental implications from petrographical and geochemical studies', *International Journal of Coal Geology* 72: 33–52 (2007)

[81] http://www.powerengineeringint.com/articles/2015/03/alstom-wins-contract-at-hungarian-coal-fired-power-plant.html

[82] *Extended Life of Hungary's Matra Power Station*, RWE Power International, Swindon, England (2008)

[83] Kuqukbayrak S., Durus B., Mericboyu A.E., Kadioglu E. 'Estimation of calorific values of Turkish lignites', *Fuel* 70: 979–981 (1991)

[84] http://www.euracoal.org/pages/layout1sp.php?idpage=475

[85] http://www.power-technology.com/projects/soma-kolin-thermal-power-plant-ankara/

[86] http://www.industrialinfo.com/news/abstract.jsp?newsitemID=233470

[87] Erik H.Y. 'Hydrocarbon generation potential and Miocene–Pliocene paleoenvironments of the Kangal Basin (Central Anatolia, Turkey)', *Journal of Asian Earth Sciences* 42: 1146–1162 (2011)

[88] http://www.euracoal.org/pages/layout1sp.php?idpage=77

[89] http://www.lahmeyer.de/projekte/energie/konventionelle-stromerzeugung/single/article/turceni-thermal-power-plant-units-3-to-6-romania.html

[90] http://bankwatch.org/news-media/for-journalists/press-releases/ebrd-suspends-loan-romanian-coal-plant-turceni

[91] http://www.hadek.com/project-reports/rovinari-power-station

[92] http://www.sourcewatch.org/index.php/Mintia-Deva_Power_Station

[93] http://www.alstom.com/press-centre/2012/9/alstom-to-supply-air-quality-control-systems-in-taiwan-and-romania/

[94] http://www.country-data.com/cgi-bin/query/r-14861.html

[95] Dimitrijevic Z., Tatic K. 'The economically acceptable scenarios for investments in desulphurization and denitrification on existing coal-fired units in Bosnia and Herzegovina', *Energy Policy* 49: 597–607 (2012)

[96] http://www.sourcewatch.org/index.php/Tuzla_Thermal_Power_Plant

[97] Dellantonio A., Fitz W.J., Custovic H., Repmann F., Schneider B.U., Grunewald H., Gruber V., Zgorelec Z., Zerem N., Carter C. Markovic M., Puschenreiter M., Wenzel W.W. 'Environmental risks of farmed and barren alkaline coal ash landfills in Tuzla, Bosnia and Herzegovina', *Environmental Pollution* 153: 677–686 (2008)

[98] Kazagic A., Smajevic I. 'Synergy effects of co-firing wooden biomass with Bosnian coal', *Energy* 34: 699–707 (2009)

[99] http://www.elektroprivreda.ba/eng/page/tpp-kakanj

[100] Thomas L., Frankland S. 'Mining at Gacko open-cast mine Bosnia-Herzegovina – a question of economics', *Geologica Belgica* 7: 267–271 (2004)

[101] http://www.eib.org/infocentre/press/news/topical_briefs/2013-march-01/tes-thermal-power-plant-sostanj-project-slovenia.htm

[102] http://www.euracoal.be/pages/layout1sp.php?idpage=80

[103] http://www.industcards.com/st-coal-slovenia.htm

[104] Slejkovec Z., Kanduc T. 'Unexpected arsenic compounds in low-rank coals', *Environmental Science and Technology* 39: 3450–3454 (2005)

[105] http://www.nama-database.org/index.php/Construction_of_a_Super-critical_Lignite_Power_Plant_TTP_Kostolac_B

[106] Životić D., Bechtel A., Sachsenhofer R., Gratzer R., Radić D., Obradović M., Stojanović

K. 'Petrological and organic geochemical properties of lignite from the Kolubara and Kostolac basins, Serbia: implication on Grindability Index', *International Journal of Coal Geology* 157: 165–183 (2015)

[107] http://wikimapia.org/6625400/T-E-Kolubara-Powerplant

[108] *RWE Identified Potential Improvements in Serbian Electricity Supply*, RWE, Swindon and Koeln (2008)

[109] http://www.tent.rs/en/pp-qmoravaq

[110] http://serbia-energy.eu/serbia-coal-mine-rembas-expands-exploration-to-new-pit/

[111] Hasania F., Shala F., Xhixha G., Xhixha M.K., Hodolli G., Kadiri S., Bylyku E., Cfarku F. 'Naturally occurring radioactive materials (NORMs) generated from lignite-fired power plants in Kosovo', *Journal of Environmental Radioactivity* 138: 156–161 (2014)

[112] *Fuel and Energy Abstracts* May 1997, p. 134

[113] http://www.industcards.com/st-coal-bosnia-montenegro.htm

[114] http://bankwatch.org/our-work/projects/pljevlja-ii-lignite-power-plant-montenegro

[115] http://bankwatch.org/campaign/coal/macedonia

[116] *Proceedings of the International Workshop in Geoenvironment and Geotechnics*, Milos, Greece, Heliotopos Conferences (2005)

[117] Richardson M. *Effects of War on the Environment: Croatia*, CRC Press (2002)

[118] King R. *The Industrial Geography of Italy*, Routledge (2015)

[119] D'Alessio A., Vergamini P., Benedetti E. 'FT-IR investigation of the structural changes of Sulcis and South Africa coals under progressive heating in vacuum', *Fuel* 79: 1215–1220 (2000)

[120] Amorino C., Bencini R., Cara R., Cinti D., Deriu G., Fandinò V., Giannelli A., Mazzotti M., Ottiger S., Pizzino L., Pini R., Quattrocchi F., Sardu R.G., Storti G., Voltattorni N. 'CO_2 geological storage by ECBM techniques in the Sulcis area (SW Sardinia Region, Italy)', *Second International Conference on Clean Coal Technologies for our Future*, Sardinia (2005). Accessible at: http://www.cct2005.it

[121] Harrison T.S., Harrison T. 'Palynology of the late Miocene *Oreopithecus*-bearing lignite from Baccinello, Italy', *Palaeogeography, Palaeoclimatology, Palaeoecology* 76: 45–65 (1989)

[122] Basilici G. 'Sedimentary facies in an extensional and deep-lacustrine depositional system: the Pliocene Tiberino Basin, Central Italy', *Sedimentary Geology* 109: 73–94 (1997)

[123] Sanjuan J., Martín-Closas C. 'Charophyte palaeoecology in the Upper Eocene of the Eastern Ebro basin (Catalonia, Spain): biostratigraphic implications', *Palaeogeography, Palaeoclimatology, Palaeoecology* 365: 247–262 (2012)

[124] White C.M., Collins L.W., Veloski G.A., Irdi G.A., Rothenberger K.S., Gray R.J., LaCount R.B., Kasrai M., Bancroft M., Brown J.R., Huggins F.E., Shah N., Huffman G.P. 'A study of Mequinenza lignite', *Energy & Fuels* 8: 155–171 (1994)

[125] http://link.springer.com/article/10.1007%2FBF00578261

[126] Néraudeau D., Allain R., Ballèvre M., Batten D.J., Buffetaut E., Colin J.P., Dabard

M.P., Daviero-Gomez V., El Albani A., Gomez B., Grosheny D., Le Loeuff J., Leprince A., Martín-Closas C., Masure E.M., Mazin J-M., Philippe M., Pouech J., Tong H., Tournepichem J.F., Vullo R. 'The Hauteriviane Barremian lignitic bone bed of Angeac (Charente, south-west France): stratigraphical, palaeobiological and palaeogeographical implications', *Cretaceous Research* 37: 1–14 (2012)

[127] Duxbury J. 'Prediction of coal pyrolysis yields from BS volatile matter and petrographic analyses', *Fuel* 76: 1337–1343 (1997)

[128] Come B., Chapman M.A. *Natural Analogues in Radioactive Waste Disposal*, Springer Science & Business Media (2012) [a Google ebook]

[129] Woodruff S.D. *Methods of Working Coal and Metal Mines: Planning and Operations, Volume 3*, Elsevier (2013) [a Google ebook]

[130] Wintle M. *An Economic and Social History of the Netherlands 1800–1920*, Cambridge University Press (2000)

CHAPTER 6
ELECTRICITY GENERATION III – NORTH AMERICA

6.1 Introduction

We are told in [1] that North Dakota and Texas are the main lignite-producing states of the USA. North Dakota has made an appearance in sections 1.3.3, 1.4, 2.3 and 3.1 of the book and 'Beulah Zap' in particular has been referred to. Although it is exceeded in lignite production by Texas, North Dakota is sometimes seen as being the USA's lignite producer par excellence. Accordingly the first part of this chapter will focus on North Dakota. Other lignite-utilising states will follow.

6.2 North Dakota (ND)

6.2.1 ND lignites

Limited petrographic information was given in Chapter 2. An 'average quality of commercially shipped' ND lignite is reviewed in [2]. This gives a value of 10 603 BTU lb^{-1} (24.7 MJ kg^{-1}) for the calorific value on a dry basis and 14.4% for the ash content, also dry basis. The sulphur content, dry basis, is given as 1.3%. Production of lignite in ND in 2014 was 28.7 million tons (26.0 million tonnes) [3]. Major power stations in ND will now be considered in turn.

6.2.2 Coal Creek power station

The largest power plant in ND, this has two 550 MW units [4,5] and takes lignite from ND's Falkirk mine (see Table 1.1) which it burns as p.f. All of the electricity produced at Coal Creek is sold to Minnesota.

One of the 550 MW units at Coal Creek began operation in 1979 and the second in 1980 [5]. The two units use between them 22 000 tons (19 954 tonnes) of lignite per day. It uses superheated (not supercritical) steam.

Coal Creek was the scene of development of the DryFining™ process [6] the concept of which is simple: heat from the turbine on fluid exit is used partially to dry the lignite before firing. The calorific value of the lignite as fired has in this way been raised from 14.5 to 16.6 MJ kg^{-1}.

6.2.3 Coyote power station and R.M. Heskett power station

These are being considered together as they both use lignite from the Beulah mine. Coyote has one 420 MW unit [7,8] and is in its 35th year of operation. The flue gases from the power station contain significant mercury [9]. To the use of fly ash in building materials previously described can be added another such use. Fly ash can be used to consolidate drill cuttings from newly created oil wells to make them easier to dispose of, and fly ash from Coyote is so used [10]. It is sold, not passed along at handling cost, to the oil field operator.

The R.M. Heskett power station in Mandan ND has two non-equivalent units providing 100 MW [11]. The two units, which were installed at times almost a decade apart, do not use the same combustion technique and neither uses p.f. One of the units uses a spreader stoker, which discharges lumps of fuel as required onto a grate for combustion. The other unit at R.M. Heskett is a fluidised bed.

6.2.4 Leland Olds and Antelope Valley power stations

The Leland Olds power station near Stanton ND uses lignite from the Freedom mine (see footnote 4), a few miles from Beulah [12]. The power station generates 656 MW from two non-equivalent units. Its environmental control measures include low-NO_x burners. Some of the electricity from Leland Olds is exported to South Dakota. The Antelope Valley power station, also using lignite from the Freedom mine, is considerably larger. It produces 900 MW from two equivalent units. Freedom mine is, in terms of annual production, the largest lignite mine in the USA and supplies several facilities in addition to the Leland Olds and Antelope Valley power stations [13].

6.2.5 Milton R. Young power station

This has a nameplate capacity of 705 MW [14] from two non-equivalent units using lignite from the mine in Center ND [15]. Reference [16] gives the calorific value of Center lignite on an as-received basis as 15.3–16.4 MJ kg^{-1}. This information together with data from [14] provides a basis for the semi-quantitative calculations shown in the box.

We use the mid value of the calorific value range, which is 15.9 MJ kg^{-1}. The power plant uses ≈ 4.5 million tons – 4.1 million tonnes – of lignite annually from Center mine to produce 705 MW. A generating efficiency will be calculated from these data as:

work out/heat in

In one year, work out $= 705 \times 10^6\ J\ s^{-1} \times (365 \times 24 \times 3600)\ s = 2.22 \times 10^{16}\ J$

In one year, heat in $= 4.1 \times 10^9\ kg \times 15.9 \times 10^6\ J\ kg^{-1} = 6.52 \times 10^{16}\ J$

Efficiency = 34%

6.2.6 Spiritwood power station

Also using ND lignite, this plant had been in mothballed status for three years by the time it came into service in November 2014 [17,18]. It has an output of 99 MW of electricity, and is a CHP facility [17].

6.2.7 Coal balls in ND

A coal ball is a consolidated inorganic structure, principally carbonate or pyrite and possibly silica, enclosing peat. The most notable examples are perhaps those found in the higher rank coal deposits of Pennsylvania. Coal balls have been identified in ND lignite beds. The coal balls from Beulah Zap were found to be entirely calcite in their inorganic part [19], and differed from those from elsewhere in ND and from coal balls generally in that plant debris is absent from the enclosed peat.

6.3 Montana (MT)

Bordering to the east with ND, MT has a much more limited programme of lignite utilisation and only one lignite mine being worked, that at Savage on the MT–ND border [20]. This services the well-known Lewis and Clark power plant in MT [21,22], to which there is reference in Chapter 3. The power station has been in existence for over 60 years, and operates at 57 MW. At the time of commencement of its operations, the Lewis and Clark power plant exceeded in capacity that of any other lignite-fired plant in the USA.

6.4 Texas (TX)

6.4.1 Introduction

Moving from states having a border with Canada to one on the Gulf Coast, we go on to consider Texan lignite, of which there is an enormous quantity and which has been mentioned in the introductory part of this book, this being an indication of its magnitude. The 2011 figure for production of lignite in Texas is 45.9 million tons (41.6 million tonnes) [23]. As with ND, lignite-fired power plants in TX will be considered in turn.

6.4.2 Petrographic comments

North Dakota and Texas are the main lignite-producing states of the USA as noted, and in [24] the maceral analyses of a group of six TX lignites and those of a group of six ND lignites are presented side by side for comparison. In four of the six TX lignites huminite was the dominant maceral group, the range being 47.3–82.0%. With the ND lignite huminite dominated in all but one of the six, and was in the range 22.6–89.5%. For the TX group the huminite reflectance was in the range 0.29–0.36%: for the ND group it was in the range 0.21–0.28%. The two groups therefore differ measurably in this regard and that the TX lignites had attained greater maturity than the set from ND, in the sense of advance along the peat-to-anthracite series, is the obvious explanation. Continuing the comparison into Hardgrove index, that for a particular TX lignite was presented in Table 1.2. Those for ND lignites as mined [16] are in the range 54.7–87.8 [16].

6.4.3 Luminant[12] Oak Grove and Sandow power plants

Oak Grove is a new facility, having begun generation and supply in 2010. It uses supercritical steam and has two units each generating at 800 MW [25]. Lignite for Oak Grove is obtained from the nearby Kosse mine [25]. In 2010, the year the Oak Grove plant started up, 6.2 million tons (5.6 million tonnes) of lignite were mined at Kosse.

Sandow power plant has been in service since 1953, when it had a single 121 MW unit. It evolved between then and 1981, by the acquisition of new units, to 950 MW [26]. In 2009 – just as Oak Grove was coming into being – a new unit was started up there which added 580 MW to the capacity making it 1534 MW.[13] Lignite for Sandow is obtained from the Three Oaks mine, which was bought by Luminant from Alcoa in 2007. The point is made in [27] that production at Kosse to service Oak Grove and increased production at Three Oaks to service the new unit at Sandow required heavily expanded activity at the two mines at the same time, and that this was achieved smoothly.

6.4.4 Luminant Big Brown power plant

This is fired by lignite which is supplemented by sub-bituminous coal from Powder River Basin.[14] Its capacity is 1150 MW. The lignite is from two mines: Big Brown and Turlington. It was announced in August 2014 [28] that Turlington will close in 2018, and its products be replaced by Powder River Basin coal. This and other curtailment on lignite-fired power generation in TX is due to an Environmental Protection Agency (EPA) rule requiring reductions in emissions [29].

Though dating only from the 1970s, Big Brown power plant is the oldest solid fuel fired power generating facility in TX [30]. It is described a little ruefully in [30] how, as state and federal emission regulations bear down more and more on such facilities as the Big Brown power plant, their viability is brought into question.

6.4.5 Luminant Martin Lake power plant

This has a capacity of 2250 MW, and here again the fuel is local lignite plus Powder River Basin coal. The lignite mines are Beckville, Liberty, Oak Hill and Tatum. Cessation of production at Beckville is imminent and this will be followed immediately by commencement of land reclamation [31]. In a research paper from 25+ years ago, at which time Beckville lignite was being used at Martin Lake as it still is at the time of going to press, its properties are listed [32]. These include a calorific value of 10 000 BTU lb^{-1} (≡ 23.3 MJ kg^{-1}) for the lignite when dry. The ash content is given as 19.4%, and the paper is largely concerned with a reduction of that by methods such as those referred to in section 2.3 of this book. Reduction to 10% was in fact reported [32]. Tatum lignite, and its use at Martin Lake, features in a relatively recent peer-reviewed paper on account of the fly ash [33], which is 48.7% by weight SiO_2. SiO_2 and Al_2O_3 together comprise 65.3%. There was very small (< 1%) 'loss on ignition' with the Tatum fly ash, indicating minor amounts of unburnt carbon.

---

12 Luminant is the biggest supplier of electricity in Texas. Not all of its power plants use lignite. The total capacity of Luminant, across a mix which includes nuclear, is > 15 GW.
13 Luminant owns only Units 4 and 5 at Sandow.
14 See Table 1.1.

Luminant started to work the Liberty mine as recently as the fall of 2014 [34] and coal from there will replace that from Beckville. It will supply 3 million tons (2.7 million tonnes) of lignite annually. With reference to the safety awards it has attracted, Oak Hill has been described as the 'most decorated' mine in East Texas [35]! One unit at Martin Lake was withdrawn from service over part of 2013–14, not because of any environmental issue but because natural gas had become cheap and coal-fired stations like Martin Lake were having to cut back to compete with gas-fired ones. A surge in natural gas prices saw a return to full capacity at Martin Lake in February 2014 [36]. Over the same period and for the same reason, one unit of the Luminant Monticello power plant (see below) was out of service.

6.4.6 Luminant Monticello power plant

This is of nameplate capacity 1880 MW, having drawn on lignite from the Thermo and Winfield mines as well as on Powder River Basin coal. There has been intermittent operation of Monticello at reduced capacity, in 2013–14 as noted above and at other times. For example, mining at Thermo ceased over the period 2011–14 [37] and selected units at Monticello were out of service over that time. Projected electricity production at Monticello for 2020, into which increased use of Powder River Basin coal has probably been factored, is 16 500 000 MW-hour [38] and it is easily confirmed by a simple calculation like the one performed in section 5.3.5 for the Amyntaio power station that this signifies production at full capacity. The expected carbon dioxide for 2020 is 16 692 200 000 kg, giving an emission factor of 1012 kg per MW-hour which, again, can be compared with information in section 5.3.5. All of the lignite-fired power stations operated by Luminant have been covered over sections 6.4.3 to 6.4.6.

6.4.7 San Miguel Electric Cooperative Inc. power plant [39]

This 400 MW plant draws on lignite from the San Miguel deposit at Christine, in southern Texas, a very remote location. Information on San Miguel lignite is given in [40], where the ash content is given as 22% (see below). The calorific value is 7497 BTU lb^{-1} ($\equiv$ 17.5 MJ kg^{-1}) and the sulphur content 2.2%.

There has been a fouling difficulty at this power plant [41]. Steam is generated by heat exchange between the hot post-combustion gases and incoming water. The fouling referred to is due to deposition of ash from the hot gases to the heat exchanger tubes at the hot gas side. Such deposition is promoted by the presence of sodium in the inorganic content of the lignite fuel. It is noted in [40] that San Miguel lignite ash is 4.8% in sodium measured as Na_2O. It is not difficult tentatively to interpret the fouling difficulty at San Miguel. In the presence of sulphur and oxygen (having regard to the fact that excess oxygen is used in such plants) the sodium content of the coal becomes sodium sulphate. This in molten form diffuses through previously unconsolidated layers of ash giving them the tenacity that is implied by 'fouling'. Such behaviour is very well documented, for example in reports of the former State Electricity Commission of Victoria, Australia.[15]

15 See section 10.1.

6.4.8 AEP[16] Southwestern Electric Power Co. power plant (the 'Pirkey power plant')

Of 721 MW capacity [42,43], this uses lignite from the South Hallsville mine. It commenced generation in 1985 and uses superheated steam. Properties of South Hallsville lignite are given in [44] as ash 9%, calorific value 23.2 MJ kg^{-1}, sulphur 1%, all on a dry basis. In 2007 (the most recent year for which the information is available) the Pirkey power plant generated 4 824 785 MW-hour of electricity. Round-the-clock operation at full capacity would, on the basis of the numerical information at the beginning of this paragraph, have raised 6 315 960 MW-hour.

6.4.9 Concluding remarks

The significant role of lignite in the energy mix for Texas, pre-eminently an oil state, is clear from the above. The discussion continues with Mississippi, another Gulf Coast state.

6.5 Mississippi (MS)

6.5.1 MS lignites

It is noted in [45] that the lignites of MS account for about 25% of the known lignite reserves of the Gulf Coast states. Details of a lignite from Panola County MS are given in [45] as the prelude to a report on an investigation of pyrolysis. That MS lignite more than most displays composition variations is commented upon in [45], and it is recorded that across eight samples from the same source the volatile matter ranged from 37.9 to 48.9%. Ash was in the range 28.9–40%.

Lignite for the power plant at Choctaw County in MS (see below) is from the Red Hills deposit, and is mined and supplied by the Mississippi Lignite Mining Company. Properties of it are given in [46].

6.5.2 Red Hills power plant, Chactow County

Entering service in 2001, this generates at 440 MW, and burns the lignite fuel in a fluidised bed [47]. Ash production annually is 607 700 tonnes [48], some of which finds application to road building.

Red Hills lignite as mined, at which stage it has a moisture content of 42%, has a calorific value of 13 MJ kg^{-1}. The ash is 11% and the sulphur content 0.4%. The low sulphur is of course a plus in utilisation terms. It is noted in [49] that the Mississippi lignite is part of a band of lignite extending across several states including Louisiana and Tennessee. The entire electricity production of the Red Hills power plant is sold to the Tennessee Valley Authority [50].

16 American Electric Power. 'Southwestern Electric Power Co.' is sometimes written SWEPCO, e.g. [41].

6.5.3 Kemper County

Part of the 'band' referred to above comprises the Kemper County MS lignite deposit. This has a calorific value in its moist state of 5290 BTU lb^{-1} ($\equiv$ 12.3 MJ kg^{-1}), 12% ash and 1% sulphur [51]. Under development at Kemper County [52,53] is an integrated gasification combined cycle (IGCC) facility which will use the local lignite to produce power at 582 MW. The principles of IGCC are not intrinsically difficult to grasp. The fuel, in the case of Kemper County lignite, is gasified and the combustible gas resulting is used to generate electricity at a gas turbine along the principles of the Brayton cycle. The exit gas from that is used to raise steam, and to generate electricity at a steam turbine along the principles of the Rankine cycle.

6.6 Louisiana (LA)

Lignite production in Louisiana began in 1985 at Dolet Hills. This was followed in 1989 by coal winning at another lignite mine in the state, namely the Oxbow mine [54]. Lignite from both goes to the Dolet Hills power plant, which is the only electricity generation facility in LA which uses lignite [55]. Consumption is up to 3.6 million tonnes of lignite annually and this provides 650 MW of electricity. The American Society of Mechanical Engineers (ASME) has, over a long period, authorised a procedure for estimation of the calorific value of a fuel not by measurements on the laboratory scale but by performance monitoring over a significant period of plant running time. This is sometimes called the boiler-as-calorimeter method and is the subject of the standard ASME PTC 34.

ASME PTC 34 would use temperature measurements at various stages of steam processing as a basis for determining the calorific value of the fuel. Herein we shall use the power generation rate as a means of determining the calorific value of the lignite as fired at Dolet Hills.[17] The calculation is shown in the box.

Using a plant efficiency of 35%, 650 MW of electricity requires (650/0.35) MW of heat = 1860 MW of heat.

3.6 million tonnes annually of the lignite is equivalent to:

3.6×10^9 kg/(24 × 3600 × 365 s)

= 114 kg s^{-1}

Calling the calorific value of the lignite Q MJ kg^{-1}:

1860 MW = Q MJ $kg^{-1} \times 114$ kg s^{-1}

↓

$Q = 16.3$ MJ kg^{-1}

which is within the range of values expected for a lignite 'as fired'.

17 The principles incorporated in ASME PTC 34 are being applied without any assertion that lignites come within the scope of the standard. One would intuitively expect it to apply to a lignite with a high degree of variation of composition.

AEP-SWEPCO is joint owner, with Cleco, of Dolet Hills. Dolet Hills is one of ten power plants which use solid fuel and, as noted, the only one which uses lignite. Of the other nine, five use sub-bituminous coal and, interestingly, four use petroleum coke ('petcoke'). Louisiana has a crude oil refining capacity of 3 million barrels per day [56], so that there should be petcoke available in large amounts is not surprising.

6.7 Arkansas (AR)

There is not, and never has been, significant power production from lignite in Arkansas. Even so there are large reserves and their utilisation is constantly being raised as a possibility. Brief coverage here, drawing mainly on [57], is therefore appropriate.

There was very limited use of lignite from Ouachita County AR for steam raising – stationary plant and locomotives – in about 1900. 'Oils', comprising pyrolysis products, were made from lignite from the same source in the first half of the 20th century and there was other very limited chemical manufacture. Projects in the 1970s and 1980s into steam raising with AR lignite never attained commercial viability. On a dry, ash-free basis AR lignite has a calorific value of 16 MJ kg^{-1}.

Of course, if AR lignite is ever developed its direct burning to make electricity is not the only possible use. With reference to the lignite in Ouachita County gasification has been looked into [58], with possible input from the Dakota Gasification Co. who feature later in this book. Montan wax production, on an experimental basis, from AR lignite features in section 15.2. That Arkansas lignite is allochthonous was noted in section 2.6. The same is believed to be true of the undeveloped lignite reserves of Kentucky [59].

6.8 Further comments

Utilisation of lignite for power generation in the USA is limited without being insignificant. Figures for 2011 are 73.6 million tonnes of lignite produced in the USA [60], compared with 450 million tonnes of bituminous coal and 462 million tonnes of sub-bituminous coal [61]. Even so expansion of power generation is evident, as already noted for example in relation to Kemper County. Lignite in Alaska is described in section 23.2.

6.9 Canada

6.9.1 Introduction

In 2012, Canada produced 10 467 550 tons (9.5 million tonnes) of lignite [62]. There are major amounts of lignite in Alberta and Saskatchewan (which has a border to the south with North Dakota, USA). There is lignite in British Columbia, but no production at this time. Far exceeding in quantity the lignite in Saskatchewan is that in Alberta, and the Province is also rich in higher rank coal. The Atikokan power plant in north-west Ontario once utilised lignite from western Canada but since 2012 has used entirely biomass [63].

6.9.2 Boundary Dam

This plant in Saskatchewan is approaching completion and will burn lignite as p.f. to produce electricity at 110 MW with accompanying CCS [64]. A recent research paper [65] concerned with lignite from Boundary Dam gives as background information a calorific value of 23.2 MJ kg^{-1} and an ash content of 15.3%, both dry basis. Also concerned with a Saskatchewan lignite was a petrographic study reported in [66]. The breakdown into maceral groups was huminite 73%, liptinite 7% and inertinite 11%. Boundary Dam was conceived and developed as an IGCC project with CCS, and the choice of lignite as fuel was on the basis of its proximity and abundance [67]. It was not chosen in preference say to natural gas for any other reason. There is more on Boundary Dam in Chapter 19.

6.9.3 Other lignite-fired power stations in Canada

A selection of these is listed in Table 6.1, and comments follow below.

Table 6.1 Selected lignite-fired power stations in Canada.

POWER STATION	DETAILS
Poplar River, Saskatchewan	Two 615 MW units utilising, when working at nameplate capacity, 488 tonnes of lignite per hour from the Poplar River mine [68]
Genesee, Alberta	Fuel from the nearby Genesee lignite mine. See main text
Shand, Saskatchewan	276 MW [73]. Lignite from the Estevan mine [74]

The Poplar River plant long pre-dates Boundary Dam, having commenced generation in 1981. The Poplar River mine is near Coronach, south-central Saskatchewan [69]. Analysis details for Poplar River lignite are given in [70]. On a dry basis these are volatiles 45.1%, fixed carbon 34.5% and ash 20.4%. The figures for Poplar mine are in [69] compared with those for Freedom mine in ND (see section 6.2.4). Genesee (next row) has three lignite-fired units the third of which, commissioned about a decade ago, was the first coal-fired generator steam turbine in Canada to use supercritical steam [71]. It is soon to be extended by two units providing between them about a gigawatt, which will use not lignite but natural gas [72].[18]

In section 4.2.3 lignites from Hambach, Germany featured, and in [75] lignites from German sources including Hambach and Canadian sources including Estevan (row 3 of the table) are compared in petrographic terms. Vitrinite reflectances were reported. For the three lignites taken from Estevan these were in the range 0.27–0.33%. The two from Hambach gave reflectances of 0.23 and 0.24%. This result was reproduced across the other lignites: the Canadian ones had a rank margin over the German. Another important result was that the German coals were higher in macerals of the vitrinite group than the Canadian ones.

18 On Saskatchewan lignite, see also Table 12.2.

6.9.4 Concluding remarks

Lignites are a contributor to the Canadian energy scene, which in some ways is different from those of other major countries by reason of the tar sands. These have long been a major source of liquid fuels in Canada but not many countries have such a resource. There is scope for discussion of equivalence or otherwise of the tar sands to the bitumen sands of Nigeria for example or even, in some respects, to the 'heavy crudes' of Venezuela. Their relevance here is that the place of lignite in Canada's energy mix is influenced by them. Lignites occurring in the Hudson Bay Lowland are discussed in section 23.2.

6.10 Mexico

In 2010, Mexico produced 8.5 million tonnes of lignite [76] and imports briquettes. Mexico also has a deposit of leonardite, one of fewer than ten leonardite deposits worldwide [77]. Leonardite [78], like lignite, has gone beyond the peat stage of coalification, but instead of undergoing the geochemical steps which, on a geological timescale, convert the peat to lignite has remained and been oxidised *in situ*. Leonardite is covered in Chapter 18.

6.11 Concluding remarks

Sustained usage of lignite in North America is clear, as is the fact that carbon sequestration is having to accompany established and new facilities for power generation from this source as explained more fully in a later chapter. The fact that internationally lignite is so widely encountered in geological exploration is discussed in Chapter 23. We conclude this chapter with a mention first of a thin lignite deposit at Ocean Beach in San Francisco [79]. There are around the world many more such deposits of lignite fuel the use of which would never be considered. There is a small deposit of lignite at Brandon, Vermont, which has attracted attention amongst geologists largely on account of its woody nature [80]. (Vermont has no coal mines and no coal-fired electricity [81].) A Brandon lignite petrographically examined [82] had a maceral analysis of 66% huminite. The dominant maceral in the huminite group was attrinite (see section 22.2). The lignite is high in macerals of the liptinite group at 32%. In analysis of an extract from the coal using toluene-acetone-methanol, aromatic and aliphatic hydrocarbons were identified. Alkanes in the extract were found to be largely unbranched, and chains comprising of odd numbers of carbons were in preponderance.

6.12 References

[1] http://www.eia.gov/todayinenergy/detail.cfm?id=2670

[2] *Quality Guidelines for Energy System Studies: Detailed Coal Specifications*, DOE/NETL-401/012111, United States Department of Energy (2012)

[3] http://eaglefordtexas.com/news/id/147336/north-dakota-coal-production-increases-4-percent-in-2014/

[4] https://www.lignite.com/mines-plants/power-plants/8465/

[5] http://www.greatriverenergy.com/makingelectricity/coal/coalcreekstation.html

[6] *DryFining™ Fuel Enhancement Process*, Great River Energy (2014)

[7] https://www.lignite.com/mines-plants/power-plants/8465/

[8] https://www.lignite.com/mines-plants/power-plants/coyote-station/

[9] Benson S.A., Laumb J.D., Crocker C.R., Pavlish J.H. SCR catalyst performance in flue gases derives from sub-bituminous and lignite coals. *Fuel Processing Technology* 86: 577–613 (2005)

[10] http://bismarcktribune.com/news/state-and-regional/coyote-getting-into-hot-water-with-coal-generated-steam/article_62b44796-d511-11e1-94ae-0019bb2963f4.html

[11] https://www.lignite.com/mines-plants/power-plants/r-m-heskett-station/

[12] https://www.lignite.com/mines-plants/power-plants/leland-olds-station/

[13] https://www.google.com.au/?gfe_rd=cr&ei=x6sQVaWRNLPu8wf5_oHwBQ&gws_rd=ssl#q=freedom+lignite+mine+north+dakota

[14] https://www.lignite.com/?id=74

[15] http://www.bnicoal.com/

[16] Schobert H.H. *Lignites of North America*, Elsevier (1995)

[17] https://www.lignite.com/mines-plants/power-plants/spiritwood-station/

[18] http://www.startribune.com/business/281121792.html

[19] Keighin C.W., Flores R.M., Rowland T. 'Occurrence and morphology of carbonate concretions in the Beulah-Zap coal bed, Williston Basin, North Dakota', *Organic Geochemistry* 24: 227–232 (1996)

[20] https://www.lignite.com/mines-plants/power-plants/lewis-clark-station/

[21] https://www.lignite.com/mines-plants/mines/savage-mine/

[22] http://www.industcards.com/st-coal-usa-mt-wy.htm

[23] Clower T.L., Reyes M. *Coal Mining and Coal-Fired Power Generation in Texas: Economic and Fiscal Impacts*, Center for Economic Development and Research, University of North Texas, Denton, TX (2013)

[24] Parkash S., Carson D., lgnasiak B. 'Petrographic composition and liquefaction behaviour of North Dakota and Texas lignites', *Fuel* 62: 627–631 (1983)

[25] http://www.powermag.com/luminants-oak-grove-power-plant-earns-powers-highest-honor/

[26] http://www.sourcewatch.org/index.php/Sandow_Station

[27] *Energy Future Holdings 2008–2012 Review*

[28] http://www.dallasnews.com/business/energy/20140829-east-texas-coal-mine-closing.ece

[29] http://www.luminant.com/news/spotlight/detail.aspx?FtID=23

[30] http://www.texastribune.org/2013/02/10/sierra-club-escalates-push-against-lumi-

nant-coal-p/

[31] http://www.news-journal.com/news/2014/sep/24/beckville-coal-mine-to-play-out-in-2015/

[32] Gollakota S.V., Lee J.M., Davies O.L. 'Process optimization of close-coupled integrated two-stage liquefaction by the use of cleaned coals', *Fuel Processing Technology* 22: 205–216 (1989)

[33] Diaz E.I., Allouche E.N., Eklund S. 'Factors affecting the suitability of fly ash as source material for geopolymers', *Fuel* 89: 992–996 (2010)

[34] http://pov.energyfutureholdings.com/2014/10/investing-rusk-county/

[35] http://dailysentinel.com/news/local/article_f1312151-eb0c-5bdf-8a2f-20a54ac12dc0.html

[36] http://www.reuters.com/article/2014/02/04/utilities-luminant-coal-idUSL2N-0L91YK20140204

[37] http://www.myssnews.com/index.php?option=com_content&view=article&id=25970:lu-minant-says-thermo-mine-will-reopen-this-summer&catid=93:local-news&Itemid=376

[38] http://enipedia.tudelft.nl/wiki/Monticello_(tx)_Powerplant

[39] http://www.kiewit.com/projects/mining/contract-mining/san-miguel-lignite-mine/

[40] Tillman D.A., Duong D.N.B., Harding N.S. *Solid Fuel Blending: Principles, Practices and Problems*, Elsevier (2012)

[41] http://www.powermag.com/clinker-minimization-at-san-miguel-electric-co-op/

[42] https://www.swepco.com/info/news/viewRelease.aspx?releaseID=169

[43] http://www.industcards.com/st-coal-usa-tx.htm

[44] Lu X., Wang T. 'Investigation of low rank coal gasification in a two-stage downdraft entrained-flow gasifier', *International Journal of Clean Coal and Energy* 3: 1–12 (2014)

[45] Farage D.L., Williford C.W., Clemmer J.E. 'Pyrolysis of Mississippi lignite in a fixed bed', *Fuel Processing Technology* 16: 35–43 (1987)

[46] Yongue R.A., Laird R. *Gasification of High-moisture Mississippi Lignite at the Power Systems Development Facility*, Southern Company Services, Wilsonville, AL (2010)

[47] http://www.usc.edu/schools/price/research/NCEID/Profiles/Mini_Sites/Red_Hills.html

[48] Hawkey G.M. 'Beneficial applications of circulating fluidised bed ash at Mississippi Lignite Mining Company's Red Hills Mine', Paper presented at the OSM Forum in the 2005 World of Coal Ash, Lexington, KY

[49] Yongue R.A., Laird R. *Gasification of High-moisture Mississippi Lignite at the Power Systems Development Facility*, Southern Company Services, Wilsonville, AL (2010)

[50] http://www.energyjustice.net/map/displayfacility-69603.htm

[51] 'CO_2 capture at the Kemper County IGCC Project', 2011 NETL CO_2 Capture Technology Meeting, Pittsburgh, PA (http://energy.nd.edu/events/2011/08/22/5613-2011-co2-capture-tech-nology-meeting/)

[52] https://sequestration.mit.edu/tools/projects/kemper.html

[53] http://www.power-technology.com/projects/kemper-county-integrated-gasification-combined-cycle-igcc-power-plant-mississippi/

[54] http://dnr.louisiana.gov/index.cfm?md=pagebuilder&tmp=home&pid=305#plants

[55] https://www.cleco.com/-/power-plants

[56] http://www.lmoga.com/industry-sectors/

[57] http://www.geology.ar.gov/energy/lignite.htm

[58] http://www.teamcamden.com/index.php/partnership-home/119-energy/57-south-arkansas-rich-in-lignite-resources

[59] Hower J.C., Rich F.J., Williams D.A., Bland A.E., Fiene F.L. 'Cretaceous and Eocene lignite deposits, Jackson Purchase, Kentucky', *International Journal of Coal Geology* 16: 239–254 (1990)

[60] https://data.un.org/Data.aspx?q=United+States+datamart%5BEDATA%5D&d=EDATA&f=cmID%3ALN%3BcrID%3A840

[61] http://www.eia.gov/totalenergy/data/annual/showtext.cfm?t=ptb0702

[62] http://knoema.com/atlas/Canada/topics/Energy/Coal/Production-of-Lignite-Coal

[63] http://www.opg.com/generating-power/thermal/stations/atikokan-station/Pages/atikokan-station.aspx

[64] http://www.globalccsinstitute.com/project/boundary-dam-integrated-carbon-capture-and-sequestration-demonstration-project

[65] Karimipour S., Gerspacher R., Gupta R., Spiteri R.J. 'Study of factors affecting syngas quality and their interactions in fluidized bed gasification of lignite coal', *Fuel* 103: 308–320 (2013)

[66] Beaton A.P., Goodarzi F., Potter J. 'The petrography, mineralogy and geochemistry of a Paleocene lignite from southern Saskatchewan, Canada', *International Journal of Coal Geology* 17: 117–148 (1991)

[67] https://sequestration.mit.edu/tools/projects/boundary_dam.html

[68] http://www.townofcoronach.com/powerplant.html

[69] http://westmoreland.com/location/poplar-river-mine-saskatchewan/

[70] Pavlish J.H., Holmes M.J., Benson S.A., Crocker C.R., Galbreath K.C. 'Application of sorbents for mercury control for utilities burning lignite coal', *Fuel Processing Technology* 85: 563–576 (2004)

[71] Watanabe S., Tani T., Takahashi M., Hidetoshi F. '495-MW Capacity Genesee Power Generating Station Phase 3: First Supercritical Pressure Coal-fired Power Plant in Canada', *Hitachi Review* 53: 109–114 (2004)

[72] http://www.marketwired.com/press-release/capital-power-announces-construction-plans-genesee-4-5-acquisition-portfolio-renewable-tsx-cpx-1974069.htm

[73] http://www.saskpower.com/our-power-future/our-electricity/our-electrical-system/

[74] http://westmoreland.com/location/estevan-mine-saskatchewan/

[75] Kalkreuth W., Steller M., Wieschenktimper I., Ganzt S. 'Petrographic and chemical char-

acterization of Canadian and German coals in relation to utilization potential. 1. Petrographic and chemical characterization of feed coals', *Fuel* 70: 683–694 (1991)

[76] http://www.factfish.com/statistic-country/mexico/lignite+brown+coal,+production

[77] http://mininglink.com.au/mine/maddingley

[78] Jones J.C., Russell N.V. *Dictionary of Energy and Fuels*, Whittles Publishing, Caithness and CRC Press, Boca Raton (2007)

[79] Carter R.M., Abbott S.T., Graham I.J., Naish T.R., Gammon P.R. 'The middle Pleistocene Merced-2 and -3 sequences from Ocean Beach, San Francisco', *Sedimentary Geology* 153: 23–41 (2002)

[80] Stout S.A., Spackman W. 'Notes on the compaction of a Florida peat and the Brandon lignite as deduced from the study of compressed wood', *International Journal of Coal Geology* 11: 247–256 (1989)

[81] http://www.sourcewatch.org/index.php?title=Vermont_and_coal

[82] Stout S.A. 'Aliphatic and aromatic triterpenoid hydrocarbons in a Tertiary angiospermous lignite', *Organic Geochemistry* 18: 51–66 (1992)

CHAPTER 7
ELECTRICITY GENERATION IV – ASIA

7.1 Introduction

Certain countries of Asia are now very strong as manufacturing bases, and for the role of lignites in that to be evaluated is an obvious approach for this chapter which will review several Asian countries one by one starting with China.

7.2 China

7.2.1 Introduction

China has a wealth of fuels of many kinds including oil, but has of late had such a moribund infrastructure that only with outside help has she been able to benefit from them. China was seen as the 'lucky country' at an international meeting on world coal at about the time of World War I, largely because of her anthracite mines. At that time there were only two ways in which coal could be utilised: by burning it or by gasifying it to make fuel gas or synthesis gas. For either of these anthracite was superior to lower rank coals. China's current reserves of lignite are estimated as 17.8 billion tonnes [1]. Vitrinite reflectance results on a suite of four Chinese lignites are reported in Chapter 2. Mongolia (an 'autonomous region' within China) has an abundance of lignite. This includes the reserves at Zhungeer [2] where 10 million tonnes per year are produced. There is no integrated coal production and power generation at Zhungeer. The lignite is sold to bodies such as the Shanghai Boiler Company which uses the lignite at a power plant that produces 600 MW. At 1.5 billion tonnes of accessible lignite, Zhungeer ranks amongst the ten largest coal deposits in the world [3]. The other parts of China where lignite occurs in significant quantities are in the north. In spite of having major resources, China imports some lignite, largely from Indonesia.

7.2.2 Examples of electricity generation in China from lignites

As in some previous parts of the book, the information is presented in a table (Table 7.1) with notes following.

Table 7.1 Selected lignite-fired power stations within China.

NAME/LOCATION OF POWER STATION	DETAILS
Liaoning Province	Test unit of 600 MW having performed as hoped for [4]
Shanghai Waigaoqiao power station	5 GW from bituminous coal [5]. A demonstration project on the same site in which lignite is used for power generation and to produce briquettes [6]
Longkou power station, Shanxi Province	Total capacity 1000 MW across four non-equivalent units [7]. Use of blends of higher rank coals with lignite at two of the units
Huolinhe Zhanute power station, Inner Mongolia	Commencement of operations at 400 MW. Producing only since 2013 [8]
Shanghai	Proposals to generate power with brown coal in the form of briquettes manufactured in Victoria, Australia [11]

The lignite-fired unit in Liaoning Province uses ultra-supercritical steam. Having been built with the help of experts from the USA, this facility belongs to China Power International [4]. The demonstration project referred to in the next row uses 150 000 tonnes per year of lignite, and superheated steam is used. The blending of black coal with brown at Longkou is a practice also followed at certain US power plants. Such blending is of increasing importance generally, and challenges facing those developing these blends include a sufficiently high calorific value and avoidance of a blend susceptible to spontaneous combustion if stockpiled [9]. At Huolinhe Zhanute power generation has only just begun, and a capacity of 5.7 GW is imminent. Lignite from Huolinguole will continue to be the sole fuel. A recent contribution to the literature [10] gives some details of this lignite resource, for example that it is 21.3% ash on a dry basis and 1.1% in sulphur. The eventual rating is expected to be 5280 MW. Six additional 660 MW units are currently being installed. A stacker 'reclaims' stockpiled coal for utilisation. In general such plant will have a capability of the order of kilotonnes per hour of coal 'reclaimed' [11].

With reference to the next row of the table, it should be noted that there is no export of briquettes from Victoria at the present time [12].

7.2.3 Concluding remarks

The point is made in [13] that the usage of lignite in China, and R&D into such usage, have both been light over the decades. Reconstruction is the keynote of China's activity and that lignite, domestic or imported, will make a major contribution to the energy requirements for that is expected. The huge investment at Huolinhe Zhanute is evidence of this. Details of a recent research investigation conclude the coverage of China in this chapter. In section 6.11 Brandon lignite from the USA was similarly dealt with though in a less detailed way.

There is lignite at Zhaotong in south-west China [14], and by drawing on a recent study of this [15] some points relevant to the carbon structure of lignites generally can be made. Tools including carbon-13 n.m.r., Fourier-transform infra-red, mass spectrometry and thermogravimetry were used in [15] to determine details of the organic structure of the lignite. The carbon content from ultimate analysis was 52.5%, dry, ash-free basis. Major findings include the fact that 42% of the carbon is present as aromatic rings, largely in pairs as in naphthalene. A difference from naphthalene is that in the lignite the hydrogens in the aromatic rings are heavily substituted. The most abundant group in the aliphatic part is methylene CH_2. The lignite is approximately 40% in oxygen, dry, ash-free basis. The oxygen atoms in the coal are largely attached to aliphatic carbons. Nitrogen (1% dry, ash-free basis) is present in structures including five-membered heterocyclic rings. (See also section 15.6.)

7.3 Japan

Japan has not mined coal of any rank since 2002 and is not currently importing any lignite [16]. As noted above, brown coal briquettes from Victoria (Australia) are not being exported at all at present because of a regime in the Latrobe Valley (see later chapter) different from the earlier one, which persisted for decades, where brown coal production was integrated with electricity production. At that time some coal was diverted to the briquette works close to one of the power stations. Prior to this recent injunction there was major export of briquettes from Victorian brown coal. Japan imports huge amounts of black coal from Australia but is currently unable to import brown coal from there for the reason given above. Even so, R&D into making pellets from Victorian brown coal for use in power generation in Japan in the longer term is under way [16].

7.4 Thailand

7.4.1 Introduction

In 2012 Thailand produced 20 million tons (18.3 million tonnes) of lignite [17]. The first use of Thai lignite in power generation was in 1953 when the mine mouth generating station at Mae Moh came into being [18]. Since then the facility has been expanded and, in around 2000, reduced as units were decommissioned. It now operates at 2400 MW.

7.4.2 Mae Moh

Reference [19] gives for Mae Moh lignite the analysis figures of 38.2% volatiles, 22.3% fixed carbon and 39.5% ash, which can be compared with the information on Mae Moh in Table 1.1. The sulphur content is around 3%. The high sulphur in particular has made for difficulties. The year 1992 is usually regarded as that in which the environmental difficulties with Mae Moh came to the notice of the world. In October that year a concentration of sulphur dioxide of 3418 μg m^{-3} was measured close to Mae Moh when the standard which applied was 1300 μg m^{-3}. These are converted to different units in the box.

1300 μg m^{-3} = $(1300 \times 10^{-6})/64$ mol = 2.0×10^{-5} mol m^{-3}

1 m^3 of any gas or gas mixture at 1 bar, 298 K contains 40 moles. The amount of sulphur dioxide on a parts per million (p.p.m.) basis is then:

$(2.5 \times 10^{-5}/40) \times 10^6 = 0.5$ p.p.m.

3418 μg m^{-3} therefore corresponds to 1.3 p.p.m.

That human beings were being exposed to > 1 p.p.m. of sulphur dioxide was a grave matter. Ambient standards of parts per hundred million are aimed for by enforcement bodies. (The figures above, appertaining to sites close to the power station, will be emission standards which are of course higher than ambient standards.) Measures were taken to implement flue gas desulphurisation (FGD) at Mae Moh. By 2001 FGD efficiencies of 90–97% were being achieved there [20]. The FGD required 446 137 MW-hour of electricity, about 3% of the 15.5 TW-hour of electricity generated at Mae Moh that same year [20]. As with other FGD operations described in this book, there was gypsum as a product.

All of the present units at Mae Moh have electrostatic precipitators achieving > 99% removal of fly ash, and all are equipped with low-NO_x burners [21]. Alstom (see sections 5.5.2 *et seq.*) are to install ultra-supercritical steam plant at Mae Moh, making it the first lignite-fired power plant in Asia to use ultra-supercritical steam [22]. This obviously signifies heavy investment in the future of Mae Moh and therefore of lignite utilisation in the region.

7.4.3 'Thai-Lao Lignite'

This term relates to lignite from Hongsa in Laos, close to the border with Thailand [23]. In the early 1990s when Thailand needed to increase her electricity production, 'Thai-Lao Lignite' was formed. The intention was that this body would build a power plant near Hongsa, electricity from which would be sold to Thailand. For legal and other reasons it was not a success. Activity at the Hongsa mine however continues (see below).

7.5 Laos and Vietnam

Laos has 500 million tonnes of known lignite reserves [24]. A power plant at Hongsa, constituted in commercial and legal terms differently from the previous one, is now being commissioned and commencement of power generation began in mid-2015 [25] and will use local lignite as fuel. By March 2016, by which time all of the three units comprising the plant will be working, nameplate capacity will be 1878 MW. Purchasers of the electricity will include the Electricity Generating Authority of Thailand (EGAT), making the electricity an export, and the local provincial authority in Laos.

Cambodia has lignite reserves [26] and also imports it in modest quantities [27]. There is no lignite-fired power station in Cambodia and only one using coal of any rank [28]. In Vietnam there is lignite at the Na Duong mine and this has been used in power generation since 2005 [29]. Its current capacity between two units, each with a fluidised bed, is 100 MW and plans by the operator Vinacomin to install another 100 MW of capacity, also with a fluidised bed, have government approval.

7.6 Malaysia

The 2012 figure for lignite production by Malaysia was nil [30]. She imports coal heavily from Indonesia and from Australia [31]. Sarawak in eastern Malaysia is productive of coals across the range of rank and this includes lignite. In Mukah in east Malaysia there is a proposed new power station [32,33] which will use locally won lignite.

7.7 Indonesia

A former member of OPEC, Indonesia has oil as well as coal, the latter having become a major product only since the 1980s. The 2010 figure for production of lignite in Indonesia was > 160 million tonnes [34]. The comment has been made [35] that mine mouth power generation with lignite, for a very long time the practice in many countries, is only just beginning in Indonesia. The most promising parts of Indonesia for development of this are Sumatra and Kalimantan, and proposals are in place. One such proposal is the AEL-Sumatra Lignite Based Thermal Power Plant [36]. In an article in the *Jakarta Post* [37] as recently as October 2014, the hope that projects like it will become a reality is strongly expressed. Indonesia has a very low *per capita* electricity consumption: 680 kW-hour, compared with 13 246 kW-hour for the USA and 10 712 kW-hour for Australia [38]. Simple calculations which the interested reader can perform for him/herself will demonstrate that to meet a significant proportion of that with lignite will not require heavy capitalisation in terms of generating capacity. This gives credence to the viability of proposed lignite-fired power generation in Indonesia.

7.8 Myanmar

An article in *The Straits Times* in 1920 [39] quotes a contributor to the *Rangoon Times* as lamenting the fact that the only known coal reserves of Burma (later to become Myanmar) were lignite, correctly described in [39] as being young coal. It goes on to mention in a very positive spirit that German lignites had shown potential as the basis of liquid fuel manufacture. This was probably with reference to the Bergius process, which as reported in section 14.3 began in 1913. A hope that this might be extended to lignites from Burma is expressed, and it is noted that by the time of the article trials with lignite from Burma as pulverised fuel in electricity generation had delivered promising results. There is only minor lignite usage in Myanmar, where a mere 13% [40,41] of the population have access to domestic electricity.

7.9 Israel and bordering countries

Arguably belonging in the chapter on European countries in view of its coast with the Mediterranean, Israel, being in western Asia, formally belongs to this part of the book. There is lignite at the Hula basin in northern Israel. It is up to 70% water when in the bed, and consequently has a calorific value as mined as low as 3 MJ kg^{-1} [42]. The ash contents are moderate at 15–20%. The reserves at Hula have been estimated as being sufficient to generate 1200 MW of electricity for 30 years [43]. Close to Hula is Lake Kinneret, the only natural freshwater lake in Israel. Its protection would be paramount were power generation at Hula to commence.

The term from geology 'carbonate platform' is fairly self-explanatory. One such in Lebanon contains a deposit of lignite [44] which is viewed as a perturbation of the basic geological milieu attributable to entry of plant debris suspended in a swamp. This point is further developed in Chapter 21. Egypt, which also has a border with Israel, is considered in section 16.2.

7.10 Concluding remarks

Reference [45] declares Thailand to be the only lignite-producing country in SE Asia. The contents of this chapter essentially confirm that, though where there is very minor use of lignite in power generation (as in Laos) or an aspiration to larger scale use (as in Indonesia) that is of interest and has therefore been recorded. The Philippines has 41 million tonnes of lignite reserves, but no significant usage as yet. Japanese lignites, which are not currently utilised, include Nakayama lignite [46], which is 37.4% volatiles on an as-received basis and 7–8% in ash. There is also lignite at Ropponmatsu in Fukuoka Prefecture. There is significant lignite within a clay formation in central Japan and its dehydroabietic acid content has been noted [47]. This substance is a constituent of 'resin acids', used in soaps and disinfectants.

The chapter began with China, and different parts of Asia followed. The parts of central Asia in the Former Soviet Union, for example Kazakhstan, will be considered in a later chapter.

7.11 References

[1] http://www.sourcewatch.org/index.php/China_and_coal

[2] *Overcoming Production Deadlock at Zhungeer Coal Mine*, RWE Power International (2008)

[3] http://www.mining-technology.com/features/feature-the-10-biggest-coal-mines-in-the-world/

[4] http://asian-power.com/project/news/china%E2%80%99s-first-lignite-fired-usc-passes-test

[5] https://www.google.com.au/?gfe_rd=cr&ei=EwLtVNXfAqfu8wem8IHoBg&gws_rd=ssl#q=shenfu+dongsheng+coalfield+bituminous

[6] http://en.kytl.com/news/news.aspx?ID=31

[7] http://www.sourcewatch.org/index.php/Huadian_Longkou_power_station

[8] http://www.sourcewatch.org/index.php/Huolinhe_Zhanute_power_station

[9] Clarke M.C. 'Low rank coal/lignite upgrading technologies', *Australian Power Technologies* October–December: 50–53 (2013)

[10] Lu J-H., Wei X-Y., Qing Y., Wang Y-H., Wen Z., Zhu Y., Gao Y-G., Zong Z-M. 'Insight into the structural features of macromolecular aromatic species in Huolinguole lignite through ruthenium ion-catalyzed oxidation', *Fuel* 128: 231–239 (2014)

[11] van Vianen T., Ottjes J., Lodewijks G. 'Simulation-based rescheduling of the stacker–reclaimer operation', *Journal of Computational Science* 10: 149–154 (2015)

[12] http://www.smh.com.au/federal-politics/political-news/200-jobs-under-threat-briquette-maker-warns-20120501-1xx40.html

[13] http://knoema.com/qbasfqc/imports-by-country-of-origin-solid-fuels-annual-data?tsId=1262400

[14] http://en.sxcoal.com/1756/DataShow.html

[15] Li Z-K., Wei X-W., Yan H-L., Zong Z-M. 'Insight into the structural features of Zhaotong lignite using multiple techniques', *Fuel* 153: 176–182 (2015)

[16] Perkins E., Tsukasaki Y., Chaffee A.L. 'Low-temperature oxidation study of brown coal and dried brown coal products', *Proceedings of the 10th Australian Coal Science Conference* Paper 5.2, Australian Institute of Energy (2013)

[17] http://knoema.com/atlas/Thailand/topics/Energy/Coal/Production-of-Lignite-Coal

[18] http://www.egat.co.th/en/index.php?option=com_content&view=article&id=36&Itemid=117

[19] Chareonpanicha M., Boonfuenga T., Limtrakul J. 'Production of aromatic hydrocarbons from Mae-Moh lignite', *Fuel Processing Technology* 79: 171– 179 (2002)

[20] Sampattagul S., Kato S., Kiatsiriroat T., Widiyanto A. 'Life cycle considerations of the flue gas desulphurization system at a lignite-fired power plant in Thailand', *International Journal of Life Cycle Assessment* 9: 387–393 (2004)

[21] http://www.scribd.com/doc/163386356/The-Mae-Moh-Plant-Review-With-EGAT#scribd

[22] http://www.alstom.com/press-centre/2015/3/alstom-to-build-the-first-ultra-supercritical-lignite-fired-power-plant-in-asia/

[23] http://www.sourcewatch.org/index.php/Hongsa_power_station

[24] Watcharejyothin M., Shrestha R.M. 'Effects of cross-border power trade between Laos and Thailand: energy security and environmental implications', *Energy Policy* 37: 1782–1792 (2009)

[25] http://www.hongsapower.com/index.php?model=cms&view=news_page&layout=news_page&id=123#ad-image-0

[26] http://metalbureau.com/Asia_News_Minerals_Of_Cambodia.html

[27] http://www.factfish.com/statistic-country/cambodia/lignite+brown+coal,+imports

[28] Shearer C., Ghio N., Myllyvirta L., Nace T. *Boom and Bust: Tracking the Global Plant Pipeline*, The Sierra Club (2015)

[29] http://finland.org.vn/public/default.aspx?contentid=197064&culture=en-US

[30] http://knoema.com/atlas/Malaysia/topics/Energy/Coal/Production-of-Lignite-Coal

[31] http://www.thestar.com.my/News/Nation/2014/11/23/Swak-holds-nearly-allof-nations-coal-reserves/

[32] http://www.sourcewatch.org/index.php/Balingian_New_power_station

[33] http://www.sarawakenergy.com.my/index.php/about-us/what-we-do/generation-portfolio

[34] http://energy.about.com/od/Coal/a/Lignite.htm

[35] http://www.asiaminer.com/news/regional-news/5287-igniting-south-sumatra-s-low-rank-coal.html#.VRYeP28cSUk

[36] http://www.marketreportsonline.com/239676.html

[37] Accessible at: http://www.thejakartapost.com/news/2014/10/06/coal-wire-sumatra-java.html

[38] http://data.worldbank.org/indicator/EG.USE.ELEC.KH.PC

[39] 'Burma lignite', *The Straits Times* 7 September 1920. Accessible online at the time of going to press.

[40] Kyaw W.W., Sukchai S., Ketjoy N., Ladpala S. 'Energy utilization and the status of sustainable energy in Union of Myanmar', *Energy Procedia* 9: 351–358 (2011)

[41] Sovacool B.K. 'Confronting energy poverty behind the bamboo curtain: a review of challenges and solutions for Myanmar (Burma)', *Energy for Sustainable Development* 17: 305–314 (2013)

[42] Kafri U., Gersh S., Dosoretz C. 'Corrections to calorific values of lignites of the Hula Basin, Israel, for contained $CaCO_3$', *Fuel* 59: 787–789 (1980)

[43] Shuval H. *Water Quality Management Under Conditions of Scarcity*, Elsevier (2012)

[44] Noujasim Clark G., Boudagher-Fadel M.K. 'The larger benthic foraminifera and stratigraphy of the upper Jurassic/lower Cretaceous of central Lebanon', *Revue de Micropaleontogie* 44: 215–232 (2001)

[45] Ewart D.L. 'South East Asia Coal Review', *World Coal* February 2003

[46] Mashimo K., Kiya K., Sato S., Tsuchiya H., Wainai T. 'Hydrogenolysis products of pyridine extract of a Japanese lignite', *Fuel* 63: 1417–1421 (1984)

[47] Sawada K., Nakamura H., Arai T., Tsukagoshi M. 'Evaluation of paleoenvironment using terpenoid biomarkers in lignites and plant fossil from the Miocene Tokiguchi Porcelain Clay Formation at the Onada mine, Tajimi, central Japan', *International Journal of Coal Geology* 107: 78–89 (2013)

CHAPTER 8
ELECTRICITY GENERATION V – THE INDIAN SUB-CONTINENT

8.1 Introduction

Discovery of lignite in Cannanore (a.k.a. Kannur), south-western India, in 1830 was mentioned in Chapter 3. There has also been reference to the Neyveli Lignite Corporation. Some significant reserves in the sub-continent are described below.

8.2 Some scenes of lignite usage in power generation in India

8.2.1 Rajasthan

These include the Barsingsar mine in the Indian state of Rajasthan, where there has been mine mouth generation by Neyveli at 250 MW since 2009 [1]. Combustion at the Barsingsar facility is by two equivalent fluidised beds. The analysis of Barsingsar lignite is given in [2] as 10.9% ash and 39.8% volatiles on an air-dried basis. The calorific value on the same basis is 21.4 MJ kg^{-1}. The sulphur, also on an air-dried basis, is 1%. These figures are auspicious for the prospects of the fairly new plant. In addition to servicing the power station there, Barsingsar lignite is sold to local companies as fuel.

The Giral power plant, also in Rajasthan, like Barsingsar generates at 250 MW [3,4]. It is owned by RVUN who have been generating electricity in that part of India since 1949, and uses local lignite. In [5] it is emphasised not only that this lignite is very high in ash but also that the mineral content is 8–10% pyrite/marcasite, each FeS_2. On combustion sulphur dioxide from this source can add to that from organic sulphur as was noted, for example, in fairly recent work involving a lignite from Australia [6]. The two units at Giral were commissioned respectively in February 2007 and December 2008. It uses 6000 tonnes of lignite a day. Lime is used in FGD at Giral. RVUN have a portfolio of several thermal power units. Giral is the only one which uses lignite.

The VS lignite plant in Rajasthan [7] generates at 135 MW using lignite from the Gurha (East) lignite mine. Very detailed information on the performance of this power generating facility is in the public domain [8]. It includes atmospheric sulphur dioxide measurements at sites known distances from the generating facility. Over a period of a month the levels were in the range 5–7 $\mu g\ m^{-3}$, equivalent to 0.002–0.003 parts per million.

JSW Energy's Barmer power plant in Rajasthan is in a different league from any of those featuring so far in this section, producing, with lignite as the sole fuel, 1080 MW from eight 135 MW units [9].[19] It obtains its lignite from the Jalipa and Kapurdi mines, the estimat-

19 It has recently been described as 'the best lignite thermal power plant in India': http://www.quora.com/Which-is-the-best-lignite-based-thermal-power-plant-in-India

ed reserves of which are 466 million tonnes [10]. The properties of Kapurdi lignite were thoroughly researched over 20 years ago [11]. The moisture content of lignite in the bed is 37–45% and the ash in the range 6–28%. The magnitude of this reserve is noted immediately above, and such variations in properties for samples at widely differing positions within the basin and at different depths are not unexpected. The samples particularly high in ash contained visible bands of clay. Such information is of course of value in production (see section 5.3.4). Huminite is the dominant maceral group in lignites from this reserve.

8.2.2 Tamil Nadu

Moving to a totally different part of India from Rajasthan, there are two lignite mines in Neyveli in Tamil Nadu. They are run by the company of the same name (the Neyveli Lignite Corporation) whose only other mine is that in Rajasthan mentioned in the previous section. 'Neyveli lignite' is described in [12] as being 8% in ash on a dry basis and of calorific value 24.9 MJ kg^{-1}. The sulphur content is given as 2.3%, and the dominant maceral group is huminite. The huminite reflectance is 0.39%. Note the enormous advantage over the Rajasthan lignites of the low ash.

Lignite-fired power production at Tamil Nadu began in 1962 at a plant which, between then and 1970, was expanded to a total capacity of 600 MW. Further expansion took place in 1996. The second plant commenced operations in 1978 and has a nameplate capacity of 1470 MW. It supplies beyond Tamil Nadu to parts of India including Pondicherry [13]. The older power plant is to be replaced by the Neyveli New Thermal Power Plant (NNTPP) which also uses lignite [14,15]. It is being commissioned at the time of going to press and uses supercritical steam in plant supplied by Alstom.

Pulverisation at this plant is by 16 beater wheel mills, manufactured by Alstom at their Indian facility [16]. A beater wheel mill, in addition to pulverising the coal, acts as a supplier of air to the fuel, and air and fuel from it go directly to the burner [17], the air, before its role in supporting the combustion, acting as a carrier gas for the p.f. An approximate calculation is shown in the box.

Having regard to the fact that supercritical steam is used at the plant the efficiency will be taken to be 0.43, so 1000 MW of power requires 2325 MW of heat.

Assigning a calorific value for the coal as fired of 12 MJ kg^{-1} the coal requirement is 195 kg s^{-1} or 700 tonnes per hour. Each of the 16 beater wheel mills is therefore producing 44 tonnes per hour of pulverised fuel.

To the very limited degree that a comparison is valid, this is at the low end of the range for Loesche mills given in section 4.2.8.

Also in Tamil Nadu is a 250 MW lignite-fired plant owned by TAQA [18]. This has been in service since 2002 and sells its electricity to the Tamil Nadu Generation and Distribution Corporation (TANGEDCO).

8.2.3 Gujurat

In western India, this region has lignite deposits which include the Vastan lignite mine block, 300 km from Mumbai. A recent contribution to the research literature which is concerned with possible underground gasification of Vastan lignite [19] gives as background information a moisture content when in the bed of 43%. The ash content is given as 4.7%. The mine is supplied by the Gujurat Industries Power Company Ltd., whose portfolio does not consist exclusively of lignite-fired facilities. It uses lignite at the Surat Lignite Power Plant (SLPP), which has four units [20]. Fluidised beds are the combustion technique; lignite enters the bed as pieces of about 10 mm size together with lime to capture the sulphur dioxide. The bed temperature is controlled so as to limit formation of NO_x. Expansion of SLPP is planned, retaining lignite as fuel [21]. Gujurat lignite as an ingredient of drilling fluid at oil wells is discussed in section 15.5.1.

A lignite from Gujurat has been examined for petrographic characteristics including gelification index (GI) [22]. As noted in section 2.2 this is determined from maceral analysis. There is no one definition in these terms; that used in [22] was:

GI = ulminite/(textinite + detrohuminite + semifusinite + inertodetrinite)

each percentage volume basis. All of these macerals are described in reference [26] of Chapter 2 (that is not so for the Greek lignite similarly evaluated in section 5.3.6). Across a set of 12 coal samples, the GI values reported in [22] were in the range 0.2–1.5. The range was widened by two outliers, and all of the samples except those had a value close to 1.0.

8.2.4 Goa

There is lignite production in Goa which is in the minerals business widely, its products including iron ore, manganese ore and bauxite. Production of lignite at Goa in June 2014 [23] was 43 lakh tons ≡ 4.3 million tons or 3.9 million tonnes.

8.2.5 The Kashmir Valley

Lignite occurs in this part of India where local reserves are believed to be 60 lakh tons (≡ 5.4 million tonnes) [24]. It is used in heating. There is a return to lignite from this region in Chapter 23.

8.2.6 Concluding remarks

It is clear from the selection above that lignite is seen as one of the fuels of the future for power generation in India. Several of the facilities discussed are undergoing replacement or expansion whilst retaining lignite as fuel. India has a *per capita* annual electricity consumption of 684 kW-hour [24], almost exactly the same as that of Indonesia. The comments made in section 7.7 apropos of Indonesia – that at that level of demand lignite can make a significant contribution – apply to India. One of the most encouraging points noted in this chapter so far is the introduction of supercritical steam to the lignite-fired power generation scene in India. This makes for efficient production and for carbon dioxide mitigation.

8.3 Pakistan

The Thar desert lignite deposit, of which the present author has written briefly elsewhere [25], contains according to a recent quantitative survey 160 billion tonnes of lignite [26]. The *per capita* annual electricity consumption in Pakistan is even lower than that of India at 449 kW-hour [27].

The Thar lignite field was discovered only in the 1980s. One can discuss power generation there only in terms of the future, though certainly the near future as electricity production from the Thar field is expected to commence in 2017 [28]. Initial production will be at 660 MW and details are as follows [29].

Lignite from the part of the mine drawn on will be of ash content just under 10% and calorific value as mined 12.6 MJ kg^{-1}. The sulphur on an as-received basis will be 1.4%. There will be two 330 MW units each with a fluidised bed. Steam will be at 175 bar and 541°C, which represents the superheated state as the temperature corresponding to saturated steam at that pressure is 355°C [30]. Rankine efficiency will be enhanced by steam re-heating, giving a value of 37% which, for sub-critical steam, is high. Ash removal from the flue gases by electrostatic precipitation will be of $>$ 99.9% efficiency. Sulphur dioxide removal from the flue gases will be $>$ 90%.

There is lignite, as well as coals of higher rank, in the part of Pakistan known as Balochistan. At present lignite from Balochistan is used to make fertilisers via synthesis gas (see section 13.3). Development into power generation is hoped for, and in [29] an extremely important point is made in favour of lignite over higher rank coals in Pakistan. The latter have to be mined from beneath the ground, and in Pakistan the record of accidents in coal mining is dismal. Lignite from Balochistan can be mined in an open cut fashion, eliminating the hazards of below-ground mining such as methane explosions (see also section 17.3). That is not of course to say that there need not be close attention to safety in open cut mining, simply that at an open cut the setting is inherently safer. In [31] the lignites of Balochistan are numbered amongst the others in Pakistan which if utilised in power generation could produce, on the estimate given in [32], 100 GW for 30 years, providing $\approx$ 900 TW-hour annually. It might be noted in conclusion that electricity consumption in Pakistan in 2011 was 90 GW-hours [33], over 60% of it from oil and natural gas. It is evident from [28] that comparisons are being made along these lines.

At Lakhra, 30 miles from Hyderabad, there is generation in a small and erratic way from lignite [34]. The rating is 150 MW although it has sometimes produced at as little as 20 MW. In [29] lignites from Lakhra are declared to be for utilisation purposes comparable to those from Thar, low ash being common to the two.

8.4 Concluding remarks

Such lignite reserves as there are in Bangladesh, formerly East Pakistan, are not seen as being of value [35]. That is not to say that there is not scope for exploration. The only coal-fired power station in Bangladesh is Barapukuria which uses locally sourced higher rank coal. Sri Lanka has one coal-fired power station with a second one under construction. Nepal has no coal-fired power stations at all, although Nepalese lignite does exist at Lukundol in the Kathmandu basin [36].

There is very limited lignite-fired electricity generation in the Indian sub-continent, but what there is is a long way from being obsolescent or moribund and, as pointed out previously, commitment to lignite usage in that part of the world is evident and expansion expected. Bhutan, a landlocked country enclosed largely by India, is discussed in Chapter 23.

8.5 References

[1] http://www.sourcewatch.org/index.php/Barsingsar_Thermal_Power_Project

[2] Selvakumaran S., Bakthavatsalam A.K. 'Effect of chemical composition of ash on sintering of lignites in Circulating Fluid Bed Combustion and successful operation of large CFBC boilers', *Applied Thermal Engineering* 85: 135–147 (2015)

[3] http://www.cctindia.org/india-power-map/article/14-giral-lignite-power-plant.html

[4] http://rvunl.com/GiralLTPP.html

[5] Srivastava S.K. 'Recovery of sulphur from very high ash fuel and fine distributed pyritic sulphur containing coal using ferric sulphate', *Fuel Processing Technology* 84: 37– 46 (2003)

[6] Yani S., Zhang D. 'An experimental study into pyrite transformation during pyrolysis of Australian lignite samples', *Fuel* 89: 1700–1708 (2010)

[7] http://www.sourcewatch.org/index.php?title=VS_Lignite_Plant

[8] *Half-yearly Report for the period October 2011 to March 2012*, V.S. Lignite Power Pvt. Ltd.

[9] http://www.jsw.in/energy/barmer-plant

[10] http://rsmm.com/blmcl.htm

[11] Mukherjee A.K., Alum M.M., Mazumdar S.K., Haque R., Gowrisankaran S. 'Physico-chemical properties and petrographic characteristics of the Kapurdi lignite deposit, Barmer Basin, Rajasthan, India', *International Journal of Coal Geology* 21: 31–44 (1992)

[12] Irdi G.A., Booher D.V., Martello D.V., Frommell E.A., Gray R.J. 'The petrography and mineralogy of two Indian coals', *Fuel* 72: 1093–1098 (1993)

[13] http://www.nlcindia.com/index.php?file_name=about_01e

[14] http://www.nlcindia.com/index.php?file_name=about_01h

[15] http://www.thehindu.com/todays-paper/tp-national/tp-tamilnadu/neyvelis-oldest-power-plant-to-close/article6181818.ece

[16] http://economictimes.indiatimes.com/alstom-india-ltd/directorsreport/companyid-3714.cms

[17] Bajric R., Zuber N., Sostakov R. 'Relations between pulverizing process parameters and beater wheel mill vibration for predictive maintenance program set-up', *Maintenance and Reliability* 16: 158–163 (2014)

[18] http://www.taqaglobal.com/our-regions/india/overview?sc_lang=en

[19] Bhaskaran S., Anuradda Ganesh A., Mahajani S., Aghalayam P., Sapru R.K. Mathur D.K. 'Comparison between two types of Indian coals for the feasibility of Underground Coal Gasification through laboratory scale experiments', *Fuel* 113: 837–843 (2013)

[20] http://www.gipcl.com/plant.php?id=CDE46A14910D102B90E20202030C33DC

[21] http://www.sourcewatch.org/index.php/Surat_Lignite_Power_Plant

[22] Singh A., Mahesh S., Singh H., Tripathi S.K.M., Singh B.D. 'Characterization of Mangrol lignite (Gujarat), India: petrography, palynology, and palynofacies', *International Journal of Coal Geology* 120: 82–94 (2013)

[23] http://goanreporter.com/mineral-production-during-june-2014/

[24] http://koausa.org/geography/chapter3.2.html

[25] Jones J.C., Russell N.V. *Dictionary of Energy and Fuels*, Whittles Publishing, Caithness and CRC Press, Boca Raton (2007)

[26] Siddiqui F.I., Pathan A.G., Ünver B., Tercan A.E., Hindistan M.A., Ertunç G., Atalay F., Suphi Ünal S., Kıllıoğlu Y. 'Lignite resource estimations and seam modelling of Thar Field, Pakistan', *International Journal of Coal Geology* 140: 84–96 (2015)

[27] http://data.worldbank.org/indicator/EG.USE.ELEC.KH.PC

[28] http://www.dawn.com/news/1084003

[29] *Thar Coal Block II Power Project Environmental and Social Impact Assessment Final Report*, Hagler Bailly, Pakistan, 21 January 2014

[30] http://www2.spiraxsarco.com/uk/resources/steam-tables/saturated-water.asp

[31] http://ummc.webs.com/apps/documents/

[32] Jaffri G., Zhang J. 'Catalytic gasification of Pakistani Lakhra and Thar lignite chars in steam gasification', *Journal of Fuel Chemistry and Technology* 37: 11–19 (2009)

[33] https://www.google.com.au/?gfe_rd=cr&ei=jp4YVdisH6ru8we2_oCACQ&gws_rd=ssl#q=pakistan++electricity++consumption+import+export+2014

[34] http://www.dawn.com/news/1172656

[35] http://www.banglapedia.org/HT/L_0114.htm

[36] Goddu S.R., Appel E., Gautam P., Oches E.A., Wehland F. 'The lacustrine section at Lukundol, Kathmandu basin, Nepal: dating and magnetic fabric aspects', *Journal of Asian Earth Sciences* 30: 73–81 (2007)

CHAPTER 9

ELECTRICITY GENERATION VI – THE FORMER SOVIET UNION

9.1 Introduction

Countries of this vast and highly varied region of the world will be examined in turn for evaluation of their power generation with lignite. The countries of the Former Soviet Union (FSU) will be covered in descending order of population. Countries not producing lignite or using it in power generation do not feature. These include the second most populous country in the FSU – the Ukraine – where in spite of large reserves lignite mining is very minor at present, and that which is produced is not used in power generation. By the same token some countries do feature which have no current lignite-fired power stations but have had in the fairly recent past, especially if published policy statements indicate an interest in revival. The most obvious example of this is Georgia.

9.2 Russia

Table 9.1 gives some details of lignite-fired power stations in Russia, and comments follow below.

Table 9.1 Lignite-fired power stations in Russia.

POWER STATION	DETAILS
Berezovskaya, eastern Siberia, Russia [1]	Now owned and operated by E.ON and uses supercritical steam. 1600 MW from two units
Nazarovo, Russia [3]	Output of 1210 MW with lignite from a deposit of the same name nearby
Krasnoyarsk [6,7]	Nameplate capacity 208 MW. Owned by the Siberian Generating Company and in service since 1979
Ryazan [8]	Nameplate capacity 2650 MW. Entered service in 1973. See also comments in the main text
Novo-Irkutsk [9]	Uses brown coal from eastern Siberia. Capacity 655 MW between several units
Shatura [11]	Originally (1920) ran on peat. The fuel mix over the years has included lignite. Now acquired by E.ON

The lignite-fired power station at Berezovskaya was acquired by E.ON in 2007. Its original commissioning straddles in time Perestroika: unit 1 came into service in 1988 and unit 2 in 1991, each producing at 800 MW. A planned unit 3 did not come to fruition. Such failure at a time which took in, for example, the August Coup is not surprising.

Berezovskaya draws fuel from the lignite deposit of the same name. This is part of the Kansk-Achinsk lignite basin. At the time that this lignite deposit came into use it was reported [2] that its calorific value in the bed was 1000–3000 kcal kg^{-1} (4–12.5 MJ kg^{-1}). In the same report the proposed building of the Berezovskaya power station was discussed, the target capacity of which – 6400 MW – had not been achieved. A plus for the Kansk-Achinsk reserve is its proximity to the Trans-Siberian railway.

The Nazarovo power plant (next row) is also in Siberia and has been in service for over 50 years. It draws on lignite from a nearby deposit of the same name, also part of the Kansk-Achinsk basin. Properties of lignite from Nazarovo are given in [4] as 5.5% ash, 49.1% volatiles and a calorific value of 13.2 MJ kg^{-1}. The coal supplying the Nazarovo power plant, and the power plant itself, are in a steppe zone comprising grass and shrubs. In a steppe zone trees are either totally absent or present only as widely separated colonies. The ecosystem of a steppe is vulnerable, and where there is thermal power generation that calls for particularly close attention to such things as ash disposal. A scene close to the Nazarovo power plant is shown in [9] and an interesting point follows from it. In spite of the very low ambient temperatures a lake close to the plant does not freeze [5]. The heat from the power station prevents it from doing so and keeps it at about +12°C.

At Krasnoyarsk (following row) only one of three installed units is currently in service. That at Ryazan supplements the primary lignite fuel with natural gas and fuel oil. The expected quantity of electricity from the plant in 2020 is 7 519 970 MW-hour and the expected carbon dioxide production 6 478 240 000 kg [10], giving an emission factor of 860 kg per MW-hour. This can be compared, for example, with those in section 5.4.2.

Since Shatura (final row of the table) began producing electricity in 1920, one can assume that its construction commenced during the reign of Tsar Nicholas II (1868–1918). It is 150 km from Moscow.

There is lignite at Sakhalin [12]. The isolation of Sakhalin makes for expensive installation and operation of infrastructure, and a lignite from Sakhalin costs much more to mine and transport than one from Siberia other things being equal [13]. A new power station there is expected to be powered by lignite [14], although there have been proposals that natural gas be used instead. Its proximity to a future liquefied natural gas (LNG) train will be of benefit to both parties. The power plant can receive natural gas from the same source as the LNG train, and can itself supply the LNG train with its considerable electricity requirements [15].

It is in some ways surprising that lignite should be used in power generation in present-day Sakhalin, which is one of the key hydrocarbon production centres of the world. Such a view does not however put things in the right perspective. The new power station replaces one which was commissioned at a time when the people of Sakhalin relied on such things as forestry, fishing and crop husbandry for their livelihoods, as some still do.

9.3 Kazakhstan

The very large coalfield at Ekibastuz in north-eastern Kazakhstan supplies the power station of the same name with coal [16]. The rank of the coal has been variously described as sub-bituminous [17], as lignite [18] and as brown coal [19]. It is surface mined and extremely high (40%) in ash. The point made at the end of section 5.5.1 on variation of rank is relevant here. Quite unequivocally lignitous in rank is the coal of the Turgay basin in central Kazakhstan where mining is expected to begin in 2020 [20] at a rate eventually sufficient to supply a 2000 MW power plant [21].

9.4 Uzbekistan

Uzbekistan has about 1.7 billion tonnes of lignite. Even so production is very low, $\approx$ 3 million tonnes annually [22]. At Angren in eastern Uzbekistan there is a power station which with lignite as fuel generates at 484 MW [23]. The lignite used is sourced from the Angren lignite mine near the national capital Tashkent [23]. The facility is set for expansion, beyond which it will continue to draw on lignite from the Angren mine and will also use higher rank coal from a different part of Uzbekistan. The chief activity at Angren is underground gasification of the lignite and this is covered in section 13.9.2.

9.5 Belarus

Belarus produces lignite which is converted to briquettes. Briquettes *per se* have a chapter later in the book. What is relevant here is production of electricity from lignite briquettes originating in Belarus. In 2011, 143 000 tonnes of lignite briquettes were used in power generation in Belarus [24]. It is straightforward to show that this would have provided of the order of 30 MW of electricity. Plans for a 400 MW lignite-fired facility in Belarus have been announced [25]. It is noted in [25] that in that region of the world small coal- or lignite-fired power generation facilities are frequently encountered. The sapropel reserves of Belarus feature in section 21.11.

9.6 Georgia

The lignite deposit at Akhaltsikhe in south-western Georgia contains > 70 million tonnes [26]. It is 36% ash, and its calorific value is given in [21] as 13.6–18.9 MJ kg^{-1}. The mine is now owned by Saknakhsiri LTD [27] who also own another lignite mine in Georgia, that at Tkibuli-Shaori. In the 20th century lignites from these mines were used *inter alia* in power production, in particular Akhaltsikhe which supplied the power station at Gardabani. At the present time coal does not feature in the electricity mix of Georgia, which is mainly hydroelectric with some natural gas. It is clear that reintroduction of lignite-fired power generation is on the agenda, including a proposed 160 MW facility at Gardabani [28].

9.7 Concluding remarks

Again, lignite usage is limited without being insignificant and its place in the future energy supply of the FSU is assured, largely because of the involvement of E.ON.

9.8 References

[1] https://www.thinkproject.com/tr/projeler/proje-ayrinti-sayfasi/project/lignite-fired-power-station-bgres-3/pa/show/?tx_tpprojectdatabase_projectdb%5Bindustry%5D=5&tx_tpproject-database_projectdb%5Bcountry%5D=0

[2] 'Soviet Union harvests Siberian coal', *New Scientist* 14 February 1985

[3] http://www.industcards.com/st-coal-russia-siberia.htm

[4] Boiko E.A., Pachkovskii S.V., Dididhin D.G. 'Experimental and numerical technique for estimating kinetic processes of thermochemical conversion of solid organic fuels', *Combustion, Explosion, and Shock Waves* 41: 47–56 (2005)

[5] http://blogs.voanews.com/photos/files/2012/02/reuters_russia_fish_farm_10Feb12-878x519.jpg

[6] http://enipedia.tudelft.nl/wiki/Krasnoyarsk-2_Powerplant

[7] http://enipedia.tudelft.nl/wiki/Krasnoyarsk-2_Powerplant

[8] http://www.ogk2.ru/eng/about/branch/ryazanskaya/

[9] http://en.irkutskenergo.ru/qa/959.2.html

[10] http://enipedia.tudelft.nl/wiki/Ryazan_Sdpp_Powerplant

[11] http://www.powerengineeringint.com/articles/print/volume-20/issue-8/features/shatura-sets-a-new-efficiency-standard-in-russia.html

[12] Nalivkin D.B. *The Geology of the U.S.S.R.*, Elsevier (2013)

[13] Minakir P.A., Freeze G.L. *The Russian Far East: An Economic Handbook*, M.E. Sharpe (1994)

[14] http://www.sourcewatch.org/index.php/Sakhalin_GRES-2_power_station

[15] http://www.downstreamtoday.com/News/ArticlePrint.aspx?aid=40097&AspxAutoDetect-CookieSupport=1

[16] http://www.sourcewatch.org/index.php/Ekibastuz-2_power_station

[17] Thomas L. *Coal Geology*, John Wiley, New York (2002)

[18] Gordon R.L. *World Coal: Economics, Policies and Prospects*, Cambridge University Press (2010)

[19] http://www.easternblocenergy.com/kazakhstan-energy-monthly/

[20] http://www.coalage.com/features/3047-kazakhstan-prepares-to-grow-coal-production.html#.VRoupG8cSUk

[21] http://www.tendersinfo.com/details/18826318

[22] http://knoema.com/EIAIES2014/international-energy-statistics-2014?tsId=1162330

[23] http://www.sourcewatch.org/index.php/Novo-Angren_power_station

[24] http://www.factfish.com/statistic-country/belarus/lignite-brown+coal+briquettes,+conversion+in+thermal+power+plants

[25] http://en.eic.in.ua/content/belarus-will-build-lignite-fired-thermal-power-station

[26] http://www.ifsdeurope.com/popup/coal.html

[27] http://www.gig.ge/en/investment-portfolio/saqnakhshiri

[28] 'Other EU Member States and Energy Community Stakeholders', *Coal Industry across Europe 2013* pp. 60–71

CHAPTER 10
ELECTRICITY GENERATION VII – AUSTRALIA

10.1 Introduction

Australia's Latrobe Valley in the region of the state of Victoria called Gippsland has been making electricity from brown coal[20] since 1924 and this has served the population of Melbourne as well as rural areas of the state over that time. Prior to that bituminous coal was brought from New South Wales for use at the City of Melbourne power station in central Melbourne. The State Electricity Commission of Victoria came into being in 1918, so responsibility for power generation was transferred to it from the City of Melbourne. There is also usage of brown coal for electricity purposes in South Australia. These will be covered in turn.

10.2 Victoria

10.2.1 Introduction

Though its beginning is outside living memory, electricity generation from Victorian brown coal is as strong as it ever was. This is partly because in recent years hydroelectricity did not perform as hoped as a result of droughts [1].

10.2.2 The coalfields and power stations

The power stations of the Latrobe Valley are described in Table 10.1.

Table 10.1 Power stations in the Latrobe Valley.

POWER STATION	DETAILS
Yallourn	Commencement of power supply from brown coal in 1924 [2]. Water content of the coal in the bed-moist state ≈ 60%. 1480 MW total capacity
Hazelwood	Coal from the Morwell open cut. Generation at 1675 MW [5]
Loy Yang	Two plants, Loy Yang A and Loy Yang B. Combined power generation 3250 MW
Anglesea (see comment in the main text re location)	150 MW from brown coal, to supply an aluminium production plant [10]
Energy Brix	Adjacent to Morwell

20 'Brown coal' is the term used in the Latrobe Valley, very rarely if at all 'lignite'. This might be due to extensive consultations with Germany during the development of power generation from brown coal in Victoria. The term 'braunkohle' was probably anglicised to 'brown coal'.

As would be expected from the period of its large-scale usage, Yallourn coal has featured in the literature frequently. Reference [3], published at a time when there had been electricity production from Yallourn coal for 25 years, gives useful details of it. The proximate analysis figures (see section 1.3.3) are given as volatiles 51.4%, fixed carbon 47.5% and ash 1.1%. The huminite maceral group is divided into six macerals. These a reader can identify from reference 26 in Chapter 2, cited there as an authoritative source for maceral classification.

Since the commencement of power generation at Yallourn noted in the table there has been decommissioning of units and installation of new ones. Now owned by Energy Australia, Yallourn now has four 1480 MW steam turbines [4] all of which use superheated steam. There is enough coal at the open cut for the power facility to deliver at its current rate until 2032 [4]. In 2013 Yallourn used 13.72 million tonnes of coal from the open cut and produced electricity in a quantity of 7.774 TW-hour.

Hazelwood power station (next row) entered service in 1964 [5]. It produces 10 TW-hour annually, and how close it operates to nameplate capacity can be calculated as:

$$\mathbf{10\ TW\text{-}hour/[1675 \times 10^{-6}\ TW \times (365 \times 24)\ hour] = 0.68}$$

so it produces at just under 70% of its rated output, which is typical. The water content of bed-moist Morwell coal is given in sources including [6], in which a value of 61% is reported. The ash content is given as 1.6% dry basis. The Hardgrove index is given in [7] as 113, a value higher than any of those in Table 1.2. Hydrogenation of Morwell coal is discussed in section 14.5.2. Hazelwood features in section 17.2.2.

Loy Yang, close to the township of Traralgon in the Latrobe Valley, is the most recent of the three brown coal fired power plants in the Valley to have come into operation having been commissioned in the 1980s. The power station draws on Loy Yang coal, use of which up to 2027 is expected [8]. A research publication from 1996 [9] – about a decade after power generation at Loy Yang began – gives the following values for the properties of Loy Yang coal. The ash is 1% dry basis, and the sulphur 0.3% dry, ash-free basis. By far the dominant maceral group is vitrinite at 91%. There is 9% exinite and a trace only of inertinite. Loy Yang coal has been investigated for non-evaporative drying by MTE and HTD (see section 1.4) [11]. HTD was found to be more effective at 300°C than at 200°C. Changes other than moisture reduction were observed, including the following: HTD is accompanied by decarboxylation; MTE and HTD both affect the pore structure, though not to the same extent.

The Victorian town of Anglesea (next row in the table) is not in the Latrobe Valley, but 170 miles west of it. The power station at Anglesea is expected to continue after the aluminium production plant which it currently serves closes down, its product becoming part of the national supply. Energy Brix (following row) is situated at Morwell, and generates power and heat for the manufacture of briquettes [12]. That there was autochthonous deposition in the formation of Victorian brown coal has been noted in a number of studies [e.g. 13,14].

The matter of overburden was raised in section 4.2.4 with a German lignite as an example. In the Latrobe Valley overburden depths are 10–20 m, atop a coal deposit in places over 200 m deep [15]. The 'overburden ratio' is therefore low. It is stated in section 4.2.6 that the overburden at lignite mines in Lusatia, Germany is 44 ± 20 m in depth.

10.3 South Australia

A recent report [16] into energy matters in South Australia contains two references to Kemper County MS, which features in Chapter 6 of this book. In the report the term 'sub-bituminous brown coal' is used to describe the coal at Leigh Creek SA. This resembles the terminology used in Kazakhstan reported in section 9.3.

In the 1950s a power station was built at Port Augusta SA. By now there are two power stations at Port Augusta which draw on Leigh Creek coal. These are described in Table 10.2.

Table 10.2 Power stations at Port Augusta using coal from Leigh Creek.

POWER STATION	DETAILS
Northern [17]	544 MW, commissioned 1985
Playford [17]	240 MW, commissioned 1963

Reserves at Leigh Creek are sufficient for a third power station of around 250 MW. Leigh Creek coals are high both in ash and in moisture, and it was because of this that activity did not begin until the 1950s as noted above. In [18] the vitrinite reflectances of three coals from Leigh Creek are presented alongside those from three coals from the black coal basin in Cranky Corner NSW. The ranges are Leigh Creek 0.43–0.49% and Cranky Corner 0.40–0.48%. The advanced rank of Leigh Creek coals can be inferred from this. The overlap of coal rank within a particular deposit is recognised in the ASTM scheme for coal classification, which also uses sub-divisions of the ranks [19]. Leigh Creek coal has been so classified as lignite A to sub-bituminous B [20]. (See also section 24.1.)

10.4 Concluding remarks

In 2012, brown coal supplied 47 555 GW-hours of electricity within Australia [21], a figure which presumably aggregates the generation in the Latrobe Valley and that in South Australia. The figure is 19.1% of the total supply, comparable to the 20.5% from natural gas. The latter is preferable in carbon dioxide emission terms. It was noted above that the operators of the South Australia power facilities using brown coal are following, with a view to adopting, developments in CCS techniques for low-rank coals, which will help brown coal maintain its place in the electricity mix. A final point is that the author has noted no mention of supercritical steam in his examination of facts and figures for this section. Introduction of that is an obvious way of increasing the acceptability in greenhouse gas terms. The author was anticipated in making this point, and has since seen it expressed in a consulting report which, being on the internet, can be assumed to be in the public domain [22]. At the time of the report there was no supercritical steam at brown coal fired power stations in Australia. There was supercritical steam at certain Australian black coal power stations by then.

Tasmania has lignite, though in small reserves widely separated from each other. In a newspaper article from 75 years ago (by which time there was major activity in Victoria, which the Tasmanians were perhaps hoping to emulate) it was reported that the isolation and the high ash precluded major utilisation [23].

10.5 References

[1] http://www.energyandresources.vic.gov.au/energy/sustainable-energy/hydroelectricity

[2] http://vhd.heritage.vic.gov.au/places/result_detail/11491?print=true

[3] Takahashi R., Sasaki M. 'Automatic maceral analysis of low-rank coal (brown coal)', *International Journal of Coal Geology* 14: 103–118 (1989)

[4] http://www.energyaustralia.com.au/about-us/what-we-do/generation-assets/yallourn-power-station

[5] http://www.gdfsuezau.com/about-us/asset/Hazelwood

[6] Bongers G.D., Jackson W.R., Woskoboenko F. 'Pressurised steam drying of Australian low-rank Coals: Part 2. Shrinkage and physical properties of steam dried coals, preparation of dried coals with very high porosity', *Fuel Processing Technology* 64: 13–23 (2000)

[7] Woskoboenko F. 'Explosibility of Victorian brown coal dust', *Proceedings of the Second Australian Coal Science Conference* pp. 118–128, Australian Institute of Energy (1986)

[8] http://www.miningoilgas.com.au/index.php/products/a-z?sobi2Task=sobi2Details&catid=12&sobi2Id=242

[9] Strugnell B., Patrick J.W. 'Rapid hydropyrolysis studies on coal and maceral concentrates', *Fuel* 75: 300–306 (1996)

[10] http://www.alcoa.com/australia/en/info_page/anglesea_coal.asp

[11] Fei Y., Artanto Y., Giroux L., Marshall M., Jackson W.R., MacPhee J.A., Charland J-P., Chaffee A.L., Allardice D.J. 'Comparison of some physico-chemical properties of Victorian lignite dewatered under non-evaporative conditions', *Fuel* 85: 1987–1991 (2006)

[12] http://www.hrl.com.au/energy-brix-australia-corporation/w1/i1001173/

[13] Verheyen T.V., Johns R.B., Bryson R.L., Malciet G.E., Blackburn D.T. 'A spectroscopic investigation of the banding or lithotypes occurring in Victorian brown coal seams', *Fuel* 63: 1629–1635 (1984)

[14] Anderson K.B., Mackay G.M. 'A review and reinterpretation of evidence concerning the origin of Victorian brown coal', *International Journal of Coal Geology* 16: 327–347 (1990)

[15] http://www.energyandresources.vic.gov.au/earth-resources/victorias-earth-resources/coal

[16] *South Australian Fuel and Technology Report*, Australian Energy Market Operator, January 2015

[17] https://alintaenergy.com.au/about-us/power-generation/flinders

[18] Li Z., Ward C.R., Gurba L.W. 'Occurrence of non-mineral inorganic elements in low-rank coal macerals as shown by electron microprobe element mapping techniques', *International Journal of Coal Geology* 70: 137–149 (2007)

[19] Standard D 388 'Standard Classification of Coals by Rank', American Society for Testing and Materials, Philadelphia

[20] Ward C.R. 'Mineral matter in Triassic and Tertiary low-rank coals from South Australia', *International Journal of Coal Geology* 20: 185–208 (1992)

[21] *Australian Energy Update*, Bureau of Resources and Energy Economics, Canberra (2014)

[22] Campisi A., Woskoboenko F. *Brown Coal R&D Scoping Study*, HRL Technology, Mulgrave, Vic. (2009)

[23] *The Advocate* (Burnie, Tasmania), 13 May 1940

CHAPTER 11
BRIQUETTES

11.1 Introduction

By 'briquette' is of course meant ground brown coal moulded into a regular shape for distribution as a general-purpose solid fuel. Sometimes a binder is used, sometimes not.

Briquette production is a major part of lignite utilisation. Use of briquettes to stabilise a p.f. flame was noted in section 5.1.3.

11.2 Briquette production by region

11.2.1 Europe and the states of the FSU

Germany is the world's largest producer of briquettes [1] having entered the business in the 19th century. The word is spelt 'brikett' in the German language.

The Knappenrode brikett factory [2] was in operation from 1928 to 1993. It had its peak production year in 1965, when it produced 1.5 million tonnes of briquettes. It was (is: it now exists as a historical building) in Lusatia where, as emphasised in Chapter 4, large amounts of lignite are produced. At Frechen there has been briquetting since 1891. The plant is now operated by RWE and produces 1 million tonnes annually of briquettes [3].

Up to the year 2000 briquettes manufactured in this region were made from lignite from the Bergheim open cut mine [4]. The mines now providing lignite feedstock for briquettes are Hambach, Garzweiler and Inden [5]. Pressures in the briquetting process are of the order of 100 MPa [6], a figure which in [6] is expressly applied to binderless briquetting of German lignites. The point is made that the drying of the raw lignite to make it suitable for the briquette press, which is accompanied by some particle shrinkage, eliminates some of the natural cohesion. The manufactured briquette owes its mechanical stability to controlled re-introduction of water and the moulding pressure.

Bulgaria has been a major producer of briquettes [7]. The briquette works at Galabovo, where there was disruption of supply in 2012 through flooding [8], uses Maritza coal. The future of the briquetting plant has been in doubt [9] for environmental reasons. Poland produced 122 000 tonnes of lignite briquettes in 1990, declining by a factor of five by 2000 [10]. There is major production at Bełchatów and some useful comments on the briquetting of lignite from this source are made in [11] in which it is noted that water removal by crushing increased the calorific value of the coal from 15.4 to 16.8 MJ kg^{-1}. In compressing the coal the proportion of the initial coal bulk volume retained as a function of applied pressure was determined for the Bełchatów lignite. For example, compression to three quarters of the original bulk volume was attained at an applied pressure of 2 kN $cm^{-2} \equiv$ 20 MPa or 200 bar. Brown coal briquettes from Hungary are manufactured at a moulding pressure of 16–17 MPa [12] and will retain their consolidation for about three months.

In the Czech Republic there has been no lignite briquette production since 2010. Coal from the Antonin seam in the Sokolov basin has been examined with a view to its briquetting in the future [13]. Results were inconclusive in terms of the viability of manufacture, though one of them is of a property not having featured in this chapter previously, the compressive strength. For the briquettes in [13] this is given as 5.6 MPa (≡ N mm^{-2}) for a briquette of width 182 mm. Semi-quantitative comparison is possible with values of the same quantity for prepared samples of a German brown coal which across diameters in the range 0.5–5 mm varies in the approximate range 27–2.5 MPa [14]. This encompasses the value for the Czech lignite briquette reported in [13]. Similarly, the compressive strength for a briquette made from a lignite originating in Turkey is given, as a function of water content, in [15]. For a briquette having experienced in production a pressure of 150 MPa the compressive stress ranges from 8 to 15 MPa across the moisture content range 8–13%. The compressive strength was also measured for briquettes made from a Turkish lignite blended with a Siberian higher rank coal. The greater the proportion of the Siberian coal the lower the compressive strength. For briquettes prepared at a pressure of 566 MPa the compressive strength was just under 30 MPa when the composition was lignite only, dropping sharply to ≈ 5 MPa for a 50:50 composition with the higher rank coal [16].

At the Rio Maior mine in Portugal [17] briquette production ceased to be competitive with imported coal.

In Russia brown coal from the Kansk-Achinsk basin is briquetted with bitumen, from oil refining residue, as a binder [18]. Not only does this raise the cost of manufacture, but the bitumen is productive of black smoke when the briquette is burnt. There have been on a development basis attempts to eliminate the bitumen in one of two ways: by using lignin as a 'milder' binder, and by blending the brown coal with a higher rank coal in such proportions that binderless briquetting becomes possible [18]. A breakdown of lignite briquette usage for Belarus in 2010 is given in Table 11.1.

Table 11.1 Breakdown of brown coal briquette usage for Belarus – consumption by different sectors, in 1000 tonnes. Information from [19].

AGRICULTURE	ENERGY SECTOR	HOUSEHOLDS	'HOUSEHOLDS AND OTHER CONSUMERS'	INDUSTRY AND CONSTRUCTION	'OTHER CONSUMERS'	'OTHER INDUSTRIES AND CONSTRUCTION'
11	0	568	704	29	125	29

The figures relate to use within Belarus. Belarus is next only to Germany as an exporter of brown coal briquettes [19]. Other Eastern bloc countries with significant exports of brown coal briquettes include Russia, the Czech Republic, Romania, Serbia and Poland. The 2010 figure for Germany was 12500 metric tonnes.

There is no lignite briquette production in Kazakhstan [20]. There are reports in the research literature (e.g. [21]) of the use of binders in briquettes made from lignites originating in Turkey. Binders tested in [21] included molasses, crude oil and bentonite, a substance widely used in drilling fluids at oil fields and inorganic (see also the final paragraph of this section). Obviously this will raise the ash content of the briquettes on burning. In the Republic of Ireland there is small-scale production of lignite briquettes and sale as domestic fuel [22]. Montan wax from Irish lignite is discussed in Chapter 15.

There is production of brown coal briquettes in Serbia [23], just over a million tonnes in 2011 [24]. In Bosnia and Herzegovina there has been briquette production from brown coal at Mostar [25], where brown coal was discovered in the mid 19th century. Elektroprivreda BiH (see section 5.8.2) generate 75 MW of electricity at Mostar [26]. In the Republic of Macedonia, lignite from the Suvodol mine has been examined closely for briquetting potential, with encouraging results [27].

The point was made above that use of an inorganic binder decreases the calorific value. It is possible however that there is an effect beyond simple dilution of the organic substance. When bentonite and cement were each used as binders in the briquetting of a Turkish lignite on a trial basis [28] the kinetic parameters for combustion were found to have been influenced by the presence of the inorganics. Less surprisingly the organic binders tested, which included molasses, also influenced the kinetics. The activation energy for oxidation of the briquettes in total absence of a binder was 42.8 kJ mol^{-1}. With the binding agents it was in the range 35.7–46.1 kJ mol^{-1}.

11.2.2 North America

On a historical note, in Texas in 1906 a 'rubbery' asphalt was being tested as a binder for briquettes from lignite [29] and compared with pitch for effectiveness. Neither would have helped towards a clean burn. In a publication from 1914 recently reprinted [30] it is stated that the briquette industry had not by then developed rapidly in the USA, and in fact it never was to. Since 2010 there has been production of lignite briquettes at the facility at South Heart ND, operated by GTL Energy [31]. It is close to a newly developed lignite open cut of the same name where there has also been interest in gasification.

11.2.3 Mongolia

As described in Chapter 7, in Mongolia lignite is used to a considerable degree in power generation. Lignite from the deposit at Baori in Inner Mongolia has been used to make briquettes. In an experimental study [32] it is reported that the briquetting of this lignite reduces the moisture content from 24 to 8%, and no binder is involved. In a further piece of work [33] briquettes from the same source are examined for water resistance, this property being correlated with oxygen functional groups including carboxyls in the lignite structure. To the content of section 7.2.3 can be added the fact that power generation from Baori lignite is planned at a target production of 1200 MW [34]. Lignite from Baganuur, Mongolia features in Chapter 13 where there is an aside on the mechanism of coalification. Lignite from Ulanqab, Mongolia features in section 17.2.3.

11.2.4 Australia

Briquetting of Victorian brown coal began at about the same time as power generation from it. German methods were followed, and Yallourn coal had the bonus of not requiring a binder. The manufacturer of the briquettes was the State Electricity Commission (SEC) of Victoria, and it comprised a significant proportion of the Commission's business.

It was noted in section 7.2.2 that no Victorian brown coal briquettes are being exported at present. In fact at the time of writing this book no Victorian brown coal briquettes are even being produced, and two briquetting plants in the Latrobe Valley are in mothballed status [35]. Introduction of the carbon tax in Australia in 2012 was seen as jeopardising the viability of briquette production and sales, and that has led to the current situation. The carbon tax has since been repealed. Production, on a regional scale, of fuel gas from Victorian brown coal briquettes is discussed in section 13.2.

11.2.5 New Zealand

The North Dakota facility [31] previously described was the scene of a production trial for briquetting of New Zealand lignite by arrangement with Solid Energy, NZ. A quantity of raw NZ lignite was taken to the plant and the trials successful in producing briquettes which later underwent combustion trials, also at the ND facility. This led to briquette production from lignite at New Zealand's New Vale mine. This became non-viable, and the state of affairs in late 2013 was that the briquettes were being returned to the mine [36].

11.2.6 India

Coal from Rajasthan featured in Chapter 8, where lignite-fired power generation in India and neighbouring countries was the topic. In [37], tests on a Rajasthan lignite for briquetting suitability with a miscellany of binders including bentonite and guar gum are reported. The findings were positive enough, but the scope of the work limited to small-scale production in rural areas. Lignite from the state of Jammu and Kashmir (see sections 8.2.5 and 23.2) is briquetted, and the products used in building heating. The Thar field in Pakistan is under development as noted in section 8.3, and briquette production concurrently with power generation is hoped for [38]. Pakistan currently imports major amounts of higher rank coal, and that this could be at least in part offset by briquettes from Thar is a sound idea.

11.3 Briquettes made from lignite plus biomass

To blend a lignite with biomass in the preparation of a briquette has the advantage not only of stretching the lignite supply but, more importantly, of introducing some carbon-neutral fuel into the briquette. The information in this section can be considered alongside that in section 20.2, where co-combustion of lignites with biomass is considered. The difference of course is that they were burnt together, not combined into a single solid fuel before burning. Table 11.2 gives details of briquettes composed of lignite plus biomass. Comments on the table follow below.

Table 11.2 Briquettes made from lignite and biomass.

REFERENCE	DETAILS
[39]	Turkish lignite blended with one of the following: pine cone, olive refuse, sawdust, paper mill waste, and cotton refuse. Briquetting pressures 50–250 MPa
[40]	Thai lignite. Rice husk and sawdust, each previously treated with sodium hydroxide, as biomass briquette constituents
[41]	Turkish lignite (from Elbistan) briquetted with hazelnut shell, with olive residue and with sunflower shell. Binders including molasses used

In the work in the first row compressive strength of the briquettes was of major interest. A value of 17.5 MPa for the lignite alone was increased to over 30 MPa with 30% sawdust or paper mill waste and to 25 MPa with 30% of pine cone. Cotton refuse had hardly any effect on compressive strength. The briquettes were in the usual sense binderless; no binder such as starch was introduced. It is believed however that the biomass acted as a binder in those briquettes in which it was present. It was noted in section 11.2.1 that lignite briquettes sometimes incorporate lignin purely as a binder. In [39] molasses were also used in blends with the lignite and these might have been expected to function as a binder. The molasses were however high in moisture as blended, and that together with the lignite's own considerable moisture content had the opposite effect and the briquettes containing molasses were not mechanically strong. Compressive strength was also the thrust of the work originating in Thailand (row 2). The briquettes were either 50:50 lignite–rice husk or 50:50 lignite–sawdust, and the compressive strength was found to depend on the time of prior 'digestion' of the biomass by sodium hydroxide solution. With rice husk as the biomass material compressive strengths of up to 30 MPa were achievable, and with sawdust values up to about 20 MPa. Elbistan (following row) also has a mention in Table 20.1. In this work, in contrast to the experience reported in [40] and summarised above, molasses were an effective binder as was starch and waste from paper pulping.

11.4 Concluding remarks

Lignite briquette production is only moderate internationally, yet this observation needs to be considered alongside sustained production at high levels by Germany and Belarus. It was noted in earlier chapters that several countries including India and parts of the FSU have capitalised in plant for lignite-fired power generation, for example by incorporating supercritical steam. That at some such places there will be a demand for concurrent briquette production is obviously possible. Briquettes are sometimes carbonised, or made from lignite which has been carbonised. A discussion of these follows in the next chapter. There is also a little more on briquetting in section 20.3.

11.5 References

[1] Mills S.J. *Global Perspective on the Use of Low Quality Coals*, IEA Clean Coal Centre (2011)

[2] http://www.pbase.com/libelletje/knappenrode_brikett

[3] http://www.rwe.com/web/cms/de/60166/frechen/

[4] Naeth J., Asmus S.C., Littke R. 'Petrographic and geophysical assessment of coal quality as related to briquetting: the Miocene lignite of the Lower Rhine Basin, Germany', *International Journal of Coal Geology* 60: 17–41 (2004)

[5] http://www.euracoal.be/pages/layout1sp.php?idpage=72

[6] Naundorf W., Wollenberg R., Schubert D. *World of Mining* 58: 32 (2006)

[7] http://www.factfish.com/statistic/lignite-brown%20coal%20briquettes,%20production

[8] http://www.novinite.com/articles/136409/Floods+in+Southern+Bulgaria+Hit+Coal,+Briquette+Production

[9] http://www.energoproekt.bg/index.php?id=158

[10] http://www.factfish.com/statistic-country/poland/lignite-brown+coal+briquettes,+production

[11] Dzik T., Marciniak-Kowalska J., Madejska L. 'Pressure agglomeration of hard and brown coals', *Chemik* 66: 445–452 (2012)

[12] Schobert H.H. *Lignites of North America*, Elsevier (1995)

[13] Buryan P., Bucko Z., Mika P. 'A complex use of the materials extracted from an open-cast lignite mine', *Archives of Mining Sciences* 59: 1107–1118 (2014)

[14] Zhong S., Baitalow F., Nikrityuk P., Gutte H., Meyer B. 'The effect of particle size on the strength parameters of German brown coal and its chars', *Fuel* 125: 200–205 (2014)

[15] Gibiiz Beker O., Kucukbayrak S. 'Briquetting of Istanbul-Kemerburgaz lignite of Turkey', *Fuel Processing Technology* 47: 111–118 (1996)

[16] Gurbuz-Beker U.I., Sadriye Kucukbayrak S., Ozer A. 'Briquetting of Afsin-Elbistan lignite', *Fuel Processing Technology* 55: 117–127 (1998)

[17] http://www.mindat.org/article.php/1301/Rio+Maior+Lignite+and+Diatomite+Mines

[18] Kuznetsova P.N., Anatoly S., Maloletnev A.S., Kolesnikova S.M. *Journal of Siberian Federal University: Engineering and Technologies* 3: 271–275 (2008)

[19] http://www.factfish.com/country/belarus

[20] http://knoema.com/EIAIES2014/international-energy-statistics-2014?tsId=1148960

[21] http://link.springer.com/article/10.1023%2FA%3A1011915829632

[22] http://www.hayesfuels.ie/wood-briquette-and-turf-products-for-sale-in-the-republic-of-ireland.php

[23] Karakosta C., Doukas H., Flouri M., Dimopoulou S., Papadopoulou A.G., Psarras J. 'Review and analysis of renewable energy perspectives in Serbia', *International Journal of Energy and Environment* 2: 71–84 (2011)

[24] http://data.un.org/Data.aspx?d=EDATA&f=cmID%3ALB%3BtrID%3A083

[25] http://connection.ebscohost.com/c/company-reports/74262698/rudnik-mrkog-uglja-mostar-doo

[26] Dimitrijevic Z., Tatic K. 'The economically acceptable scenarios for investments in desulphurization and denitrification on existing coal-fired units in Bosnia and Herzegovina', *Energy Policy* 49: 597–607 (2012)

[27] https://inis.iaea.org/search/search.aspx?orig_q=RN:31005900

[28] Altun N.E., Hicyilmaz C., Bagci A.S. 'Combustion characteristics of coal briquettes. 1. Thermal features', *Energy & Fuels* 17: 1266–1276 (2003)

[29] Professional Paper – United States Geological Survey (1906)

[30] http://www.forgottenbooks.com/readbook_text/Conservation_of_Coal_in_Canada_With_Notes_on_the_Principal_Coal_Mines_1000690640/35

[31] https://www.lignite.com/news-events-public/news-releases/gtl-energyr-squos-clean-coal-technology-proves-successful/

[32] Mo C., Shifeng Z., Yanyan Y.I., Yunpeng D. 'Characteristics of pyrolysis products of Bori [sic] lignite briquette', *Energy Procedia* 16: 307–313 (2012)

[33] http://zh.scientific.net/AMR.560-561.550

[34] *Typical Projects*, China Engineering Co. Ltd.

[35] http://environmentvictoria.org.au/media/latrobe-valley-brown-coal-plant-mothballed#.VSRi7W8cSUk

[36] https://coalactionnetworkaotearoa.wordpress.com/2013/10/15/briquette-mothballed/

[37] Mishra S.L., Sharma S.K., Agarwal R. 'Briquetting of lignite for domestic fuel', *Journal of Scientific and Industrial Research* 59: 413–416 (2000)

[38] http://en.crrssh.com/wysc/28576.html

[39] Yaman S., Sahan M., Haykiri-Acma H., Sesen K., Kucukbayrak S. 'Fuel briquettes from biomass–lignite blends', *Fuel Processing Technology* 72: 1–8 (2001)

[40] Chaiklangmuang S., Supa S., Kaewpet P. 'Development of fuel briquettes from biomass–lignite blends', *Chiang Mai Journal of Science* 35: 43–50 (2008)

CHAPTER 12
CARBONISED PRODUCTS

12.1 Introduction

Any coal on thermal decomposition will yield three classes of product: solid (coke or char), liquid (tars and oils) and gaseous. The gaseous product will be combustible with a calorific value about a third that of natural gas. This gas when manufactured on an industrial scale is called 'retort coal gas'.

One of the most widely known facts about lignites is that they do not, on decomposing, leave a hard product like coke but a powdery one called a char. By contrast a coking coal produces on decomposition a solid which is fused and hard and therefore suitable for use as a metallurgical reductant. Note that whilst it is true that all coking coals are bituminous in rank, vice versa is not true: not all bituminous coals produce on carbonisation a fused solid product suitable for use in a blast furnace. There is a time-honoured test for determining the propensity to coking of a particular bituminous coal. It is called the Gray–King assay and has become an ISO standard [1] (see Chapter 24). The standard would never be applied to a lignite, which would be outside its scope. Ion exchange sometimes precedes carbonisation of a lignite. A point of semantics is that 'carbonisation' means enriching the solid substance in carbon to the loss of oxygen in the tars and gases. These might however in a particular application be of more interest than the residual solid, and the term 'pyrolysis' is really preferable. If the desired product is a solid its likely use is as an adsorbent in which case the most important properties are the internal surface area and the porosity.

12.2 Some principles of lignite carbonisation

This will take the form of information from selected literature in tabular form (Table 12.1), accompanied by comments.

With reference to the tar from Neyveli lignite in the first row, we first note that it is an example of 'carbonisation' where the primary product is liquid. We further note that if the empirical formula of the tar was CH_2 it would correspond to $C_{29}H_{58}$. Its molar mass of about 400 g is much lower than that of asphaltenes from crude oil, which have molar masses of typically 1500 g. We note in passing that in the 1970s there were several flash pyrolysis projects into lignites, that is, rapid pyrolysis the liquid products of which could be assessed for conversion to liquid fuels.

Table 12.1 Carbonisation studies of lignites.

REFERENCE AND THEME	DETAILS
Lignite from Neyveli mine in Tamil Nadu. Tar from pyrolysis at 650°C refined to obtain wax [2]	Solvent extraction of wax from the tar product of pyrolysis. Wax of melting point 62–64°C and corresponding to a molar weight of 399 g
Kansk-Achinsk lignite pyrolysed in a fluidised bed with catalysts derived from slag [3]	Product required a carbon residue of good porosity and internal surface area
Lignites from Montana and North Dakota pyrolysed at 808°C in a flow furnace [4]	Char internal surface areas up to ≡ 500 m^2 g^{-1} obtained
Pyrolysis of briquetted lignite [5]	Required product a metallurgical reductant
Briquetting and carbonisation of coal from Tamil Nadu [6]	An activity of Neyveli
Lignite from Konin (see section 5.2.4) [7]	Carbonisation products obtained at 500–700°C reacted with urea at 350°C

By contrast with the work on the Neyveli lignite, in that on the Kansk-Achinsk lignite (following row) it was the solid product that was of interest. The emphasis is on the cheapness (see also Table 14.2). No commercial catalysts were used only slags, for example slag from a blast furnace and slag from the combustion in a furnace of Kansk-Achinsk lignite itself. Silica, alumina and lime were dominant constituents of the slags. The resulting carbon residues had internal surface areas of up to 419 m^2 g^{-1} and porosities up to 0.61 cm^3 g^{-1} making them quite suitable for use as adsorbent carbons. The internal surface areas of the chars from the coals from North Dakota and Montana in the following row were dependent on pyrolysis time, which could be controlled by rate of passage of particles through the reactor. Pyrolysis of briquettes, alluded to at the end of the previous chapter, features in the final two rows. The most common 'metallurgical reductant' (antepenultimate row) is coke from black coal, and a substitute for it has to have the hardness of coke. The idea of reacting a carbonised lignite with urea (final row) was raising of its nitrogen content. The findings were that the greatest surface area was obtained by carbonisation, activation and urea treatment rather than by urea treatment preceding activation. Internal surface areas of 3000 m^2 g^{-1} were obtained by this means, and the area was largely in the micropores.

12.3 Uses of lignite char

12.3.1 Introduction and background

In the previous section an example of adsorbent carbon production from lignite was examined. There the adsorbent carbon was a direct product of fluidised bed pyrolysis. More often, the char so obtained has to be activated for use as an adsorbent carbon. The large surface area is achieved by steam (a 'steam activated carbon') or by treatment with a chemical agent such as phosphoric acid (a 'chemically activated carbon'). An activated carbon can have an internal pore surface area as high as 1000 $m^2 g^{-1}$ and this can be measured by adsorption of a gas or vapour by the carbon being so evaluated. The two most obvious applications of activated carbons are as a decolourising agent in the chemical industry and in air purification. There is also industrial application, for example removal of particular constituents such as SO_2 from flue gas. As background to what follows, pores within an activated carbon come in three ranges of size: micropores, mesopores and macropores, in ascending order of pore diameter. The performance of an activated carbon in a particular application depends not only on the total pore volume but also on the distribution of the pores across the three categories of pore size. The pore size distribution in a particular activated carbon can be routinely determined by mercury porosimetry. Instruments are available which will measure surface area and pore size distribution simultaneously.

12.3.2 Lignite-derived activated carbons

Table 12.2, which is accompanied by comments, gives examples. Surface areas of activated carbons such as that given in the first row of the table are measured by the Brunauer, Emmett and Teller (BET) method, details of which can be found in any standard text on physical chemistry, and are called 'BET areas'.

Table 12.2 Applications of lignite-derived activated carbons.

APPLICATION AND REFERENCE	DETAILS
Activated carbon from a Turkish lignite used to decolourise an aqueous product [8]	Removal of crystal violet dye by the activated carbon, which had a surface area of 921 m^2 g^{-1}
Activated carbon from a ND lignite used in a trial study of water purification [9]	Positive results for removal of organic substances from previously untreated water by the activated carbon derived from ND lignite
Activated carbons from TX and ND lignites evaluated in terms of mercury removal from flue gas [10]	Activated carbon from TX lignite effective in elemental mercury removal at temperatures in the range 100–150°C. Mercury as sulphates and/or nitrates, from reactions with SO^x and NO^x, believed to have been present on the surface of the activated carbon from the ND lignite
Toluene adsorption from water by activated carbons from ND lignite [11]. BET areas in the range 245–370 m^2 g^{-1}. Varying amounts of cations (Na^+, Ca^{2+}) in the activated carbons	Toluene adsorption promoted by high Ca^{2+} and low Na^+, and by low Na^+ and low Ca^{2+} Toluene was chosen to represent disinfection by-products in drinking water
Chemically activated adsorbent carbons from two Polish brown coals [13]	See comments in the main text
German lignite HOK® carbonised [14]	A range of products for water and gas cleaning
Saskatchewan lignite examined for SO2 adsorption in untreated, carbonised and carbonised and steam activated forms [15]	See comments in the main text
Baori lignite (see section 11.2.3) carbonised in the temperature range 400–1000°C [16]	Rise in calorific value from 22.9 MJ kg^{-1} for the dry lignite to 28.6 MJ kg^{-1} for a char prepared at 1000°C

In [8] (first row of the table) it is noted that the adsorption process is endothermic. The activated carbon from the ND lignite in row 3 had a BET area of 245 m^2 g^{-1}, a low value. This is attributed [10] to loss of micropores in the activation process with steam. In the entry in the next row, the dependence on cation amount is believed to be due to the effects of these on surface area after carbonisation. Calgon F400 [12], a commercial adsorbent carbon made from bituminous coal, was used as a benchmark in [11].

One of the brown coals in the Polish study in the next row was from the Konin mine (see section 5.2.4). The activated carbons were prepared in a process which involved several steps: ammoxidation of the starting material; carbonisation; chemical activation with potassium hydroxide; further ammoxidation. 'Ammoxidation' simultaneously oxidises the carbon and adds to its nitrogen content.[21] Moving to the next and final row of the table, HOK means 'Herdofenkoks', rotary-hearth furnace coke. In the German language 'koks' means any solid carbonisation product of coal, whether a coke or a char. So 'braunkohle koks' means char from brown coal. The contents of the following row are concerned with SO_2 adsorption by a Saskatchewan lignite in three forms as noted. The performance is summed up below:

Untreated lignite	15 mg SO_2 per g of lignite
Lignite char	26 mg SO_2 per g char. 17 mg per g of original lignite
Steam activated char	93 mg SO_2 per g of steam activated product

A related simple calculation is given in the box.

On the hypothetical basis that in the activated carbon the adsorbed sulphur dioxide is present as a monolayer and using a surface area of 0.35×10^{-18} m^2 for a single sulphur dioxide molecule [17], the internal area of 1 g of the steam activated char is:

Avogadro's number

$(93 \times 10^{-3}$ g/64 g mol$^{-1}) \times 6.02 \times 10^{23}$ mol$^{-1} \times 0.35 \times 10^{-18}$ m$^2 = 305$ m^2

The result is lower than expected by a factor of three, which probably reflects the assumption of a monolayer. However, on the same basis the lignite without carbonisation has an internal surface area of about 50 m^2 g^{-1} which intuitively seems about right. Quite possibly the monolayer assumption is correct for the lignite but not for the activated product with its much more developed pore structure and the loss of oxygen atoms which carbonisation involves.

In the work on the Baori lignite (following row) the chars were for use in slurries as liquid fuels, so the calorific value was of major importance. The surface area (measured by BET) was highest for the char prepared at 800°C, having a value of 216 m^2 g^{-1}. The decline to 135 m^2 g^{-1} for the char prepared at 1000°C is attributed to blockage of pores and consequent loss of access to part of the surface area, a previously well-documented effect.

21 Ammoxidated carbons are being investigated for carbon dioxide capture, e.g. https://www.infona.pl/resource/bwmeta1.element.elsevier-ee63c95d-6568-3d38-a494-febce98dc6a6

Lignites then are a major feedstock for the manufacture of activated carbons via the chars which they form on heating. In terms of the organisation of this text two further points have to be made at this stage. One is that there is also a gaseous product when a brown coal is carbonised, and this will feature in the chapter on gasification. The other is that chars can be gasified with air or with steam, and this too will be discussed in the gasification chapter.

12.4 Hard chars

Lignites under most conditions of carbonisation produce a char which, almost by definition, is powdery and lacking the mechanical strength for metallurgical use. This has naturally led to R&D into making a hard, lumpy char from lignites, especially in places lacking coking coal yet requiring a carbon substance for the processing of metal ores. Such R&D into Victorian brown coals is reported in [18] in which it is acknowledged that hard chars had, long before the work described in [18], been made from Victorian brown coal. Though these did find important uses in applications including 'Heat Beads®', they were not suitable as a replacement for coke in blast furnaces. This was due not to lack of mechanical strength as with a powdery char but to difficulties with control: in a blast furnace these chars react with the metal ore much more rapidly than coke does.

The R&D in [18] is therefore directed at production of a char mechanically strong enough for use as a metallurgical reductant and having the same reactivity towards the ore as coke from a higher rank coal. The brown coal was in powdered form derived either from previously untreated coal or powdered briquettes. In either case a tar binder was used to make pellets for the carbonisation, which was under nitrogen and at a programmed temperature eventually rising to 950°C. Carbonisation times at that ceiling temperature were either 2 hours or 5 hours. Carbonised product from the unbriquetted coals gave better compressive strengths, depending on the amount of tar binder and the carbonisation time, up to 60 MPa. This very high value can be compared with the compressive strengths of lignite briquettes from various countries given in the previous chapter.

12.5 Char briquettes

The attraction of such briquettes is that, having been denuded of most of their volatiles, they burn with less smoke than coal not having been carbonised. Manufacture of char briquettes is surprisingly limited at present, and there is an obvious reason for that. Being coal derived they are not carbon-neutral, unlike chars prepared from pyrolysis of biomass which are of course carbon-neutral. Coal char briquettes do require a binder, starch being a common choice.

In [19] chars (called 'semicokes'), from the carbonisation across a range of temperatures of a Turkish lignite and a Turkish sub-bituminous coal, were combined with coal tar pitch. The interesting point is made that in carbonisation inert macerals help bind reactive ones, so in this work the chars simulate the inert macerals and the pitch the reactive ones. Discs of the pitch were placed on top of pieces of the char, and the combined pair of substances placed in an oven at temperatures in the range 60–140°C. This led to softening of the pitch and entry into the char structure. Final products of up to about 5% by weight in pitch were obtained, there being a dependence of the pitch uptake on the carbonisation temperature of the char. This dependence differed between the lignite and the sub-bituminous coal.

12.6 Char combustion

There are in the literature many accounts of lignite char combustion. Table 12.3 gives some selected examples.

Table 12.3 Summary of selected literature on lignite char combustion.

COMBUSTION CONDITIONS AND REFERENCE	DETAILS
Chars from 25 Turkish lignites in original and demineralised form studied by thermogravimetry [20]. Determination of mineral content by ISO-602 [21]	The minerals found to have a strong positive effect on combustion rates
Millimetre-size particles of Beulah Zap lignite burnt, and volatiles combustion and char combustion observed separately. A range of air pressures and temperatures [22]	'Oxidation time' defined and measured as the interval between cessation of volatiles combustion and total burnout. Oxidation times in the range 188–446 s
Lignite from Kangal, Turkey, in demineralised form, burnt. Time resolution of volatiles and char burning [23]	At 1073 K, volatiles burnout time 75.5 s. Char burnout time 922 s
Chars from two Spanish lignites carbonised at 900°C. Reaction rates in air and in carbon dioxide compared [24]	See comments in the main text

In the work described in the first row, the effect of the minerals in the original lignite on the combustion rate is marked. As a typical example, a char from a particular lignite showed an ignition temperature of 759 K whilst one from the same lignite in demineralised form having been prepared under the same carbonisation conditions showed an ignition temperature of 774 K. This trend was, with one exception, consistent for the char pairs so studied, 25 in all. In the work summarised in the following row the time for volatiles combustion was always much shorter than that for char oxidation. For the trial where the 'oxidation time' was 188 s the time for volatiles combustion was 19 s. The oxidation time of 446 s corresponds to a volatiles combustion time of 24 s. The study was of chars *per se*, not of coal combustion. The initial envelopment of the char by volatiles and their rapid combustion is alluded to in section 1.3.3. The same trend – rapid volatiles burnout and much more sluggish char burnout – is evident in the results in the next row on Kangal lignite. Kangal lignite actually features in section 5.6, where it was noted that it is hypautochthonous. This is probably the reason demineralisation was required before burning as noted in the table.

Hypautochthonous coals sometimes are high in ash, and when this is so it is due to inorganic substances conveyed by the water which 'tumbles' the plant deposition (see section 2.6).

The Spanish work in the following row compares the reactivity of the chars in air and in carbon dioxide. Reaction in carbon dioxide is of course relevant to gasification:

$$CO_2 + C \rightarrow 2CO$$

Reaction rates in air were, in mass loss terms, about half an order of magnitude higher than those in carbon dioxide. Hydrogenation of Spanish lignites is discussed in section 14.5.2. Production of chemicals from a Spanish lignite by supercritical extraction is discussed in section 15.3.2.

12.7 Concluding remarks

As already pointed out, formation of a char is always accompanied by gas and liquids (tars and oils). The former features in the part of the book on gasification which follows.

12.8 References

[1] ISO/FDIS[22] 502: Coal, determination of caking power. Gray–King coke test.

[2] Tiwari K.K., Banerji S.N., Bandopadhyay A.K., Bhattacharya R.N. 'Recovery of wax from Neyveli lignite tar and its structural characterization', *Fuel* 74: 517–521 (1995)

[3] Shchipko M.L., Kuznetsov B.N. 'Catalytic pyrolysis of Kansk-Achinsk lignite for production of porous carbon materials', *Fuel* 74: 751–755 (1995)

[4] Nsakala N.Y., Essenhigh R.H., Walker P.L. 'Characteristics of chars produced from lignites by pyrolysis at 808°C following rapid heating', *Fuel* 57: 605–611 (1978)

[5] Gupta R.C. *Theory and Laboratory Experiments in Ferrous Metallurgy*, PHI Learning Pvt. Ltd (2010)

[6] http://cuddalore.tn.nic.in/neyveli.htm

[7] Pietrzak R., Wachowska H., Nowicki P. 'Preparation of nitrogen-enriched activated carbons from brown coal', *Energy & Fuels* 20: 1275–1280 (2006)

[8] http://connection.ebscohost.com/c/articles/82583802/adsorption-crystal-violet-from-aqueous-solution-activated-carbon-derived-from-g-lba-lignite

[9] Stepan D.J. *et al. Powdered Activated Carbon from North Dakota Lignite: An Option for Disinfection By-Product Control in Water Treatment Plants*, Energy & Environmental Research Center (1995)

[10] Olsen E.S., Laumb J.D., Benson S.A., Dunham G.E., Sharma R.K., Miller S.J., Pavlish J.H. 'Comparison of the mercury flue gas-sorbent interactions on carbons from ND and TX lignites', *ACS Fuel Chemistry Division Preprints* 47: 482 (2002)

[11] Olson E.S., Sharma R.S., Eylands K.E., Stepan D.J. 'Effects of lignite cation content on

22 'Final Draft of International Standard'.

the performance of activated carbon products', *ACS Fuel Chemistry Division Preprints* 47: 473 (2002)

[12] *Filtrasorb® 400 Granular activated carbon*, Product Bulletin, Calgon Carbon, Pittsburgh, PA

[13] Pietrzak R., Nowicki P., Wachowska H. 'Ammoxidized active carbons as adsorbents for pollution from liquid and gas phases', *Polish Journal of Environmental Studies* 19: 449–452 (2010)

[14] http://www.prnewswire.co.uk/news-releases/hok-activated-lignite-efficient-filter-technology-for-environmental-sector-successful-around-the-world-152937025.html

[15] Chattopadhyaya G., Macdonald D.G., Bakhshi N.N., Soltan Mohammadzadeh J.S., Dalai A.K. 'Adsorptive removal of sulphur dioxide by Saskatchewan lignite and its derivatives', *Fuel* 85: 1803–1810 (2006)

[16] Li Y., Wang Z-H., Huang Z-Y., Liu J-Z., Zhou J-H., Cen K-F. 'Effect of pyrolysis temperature on lignite char properties and slurrying ability', *Fuel Processing Technology* 134: 52–58 (2015)

[17] http://pubchem.ncbi.nlm.nih.gov/compound/sulfur_dioxide

[18] Mamun Mollah M., Jackson W.R., Marshall M., Chaffee A.L. 'An attempt to produce blast furnace coke from Victorian brown coal', *Fuel* 148: 104–111 (2015)

[19] Arslan V. 'Investigation of bonding mechanism of coking on semi-coke from lignite with pitch and tar', *Energy & Fuels* 20: 2137–2141 (2006)

[20] Haykiri-Acma H., Ersoy-Mericboyu A., Kucukbayrak S. 'Effect of mineral matter on the reactivity of lignite chars', *Energy Conversion and Management* 42: 11–20 (2001)

[21] http://www.iso.org/iso/catalogue_detail.htm?csnumber=4705

[22] Bateman K.J., Germane G.J., Smoot L.D., Blackham A.U., Eatough C.N. 'Effect of pressure on oxidation rate of millimetre-sized char particles', *Fuel* 74: 1466–1474 (1995)

[23] Yilgin M., Pehlivan D. 'Volatiles and char combustion rates of demineralised lignite and wood blends', *Applied Energy* 86: 1179–1186 (2009)

[24] Olivella M.A., de las Heras F.X.C. 'Study of the reactivities of chars from sulphur rich Spanish coals', *Thermochimica Acta* 385: 171–175 (2002)

CHAPTER 13
GASIFICATION

13.1 Introduction

Gasification of coals was taking place in the 19th century, when the required product was fuel gas for supply to homes and businesses, often at that time –and more recently – referred to as town gas. There were and are a number of ways of producing such a gas, one of which was to decompose coal in a retort. This gas would be accompanied by by-products, a carbonised solid and tars/oils. Similarly where the primary product was coke as a metallurgical reductant, the by-product gas, called coke-oven gas, would be suitable for use as town gas and this happened in places including Pittsburgh. The preceding two sentences relate to bituminous coal. In the next section we shall examine town gas production from lignites.

13.2 Town gas

In most parts of the developed world this has been replaced with natural gas. Even so a discourse on gasification of lignite requires that town gas production be given its due place and accordingly two examples follow (see Table 13.1).

Table 13.1 Examples of town gas from lignite.

LOCATION	DETAILS
Boulder CO [1]	In 1902, the first utility to make town gas from lignite
Victoria, Australia	Production of town gas from brown coal briquettes from 1956 to 1969

It is recorded in [1] that in about 1920 the plant at Boulder took on an extra process which was by then common at gasification plants for higher rank coals. Gas obtained by reacting steam with lignite gives a product the calorific value of which is recorded in [1] as 300–350 BTU per cubic foot. Taking the middle of the range of 325 BTU per cubic foot, this is equivalent to 12 MJ m^{-3}. The gas was then 'carburetted' by vaporising oil from petroleum and blending that with the gas, and was raised in calorific value to about 20 MJ m^{-3}.

This is a suitable point at which to discuss Colorado lignites. They occur in the Denver basin, which extends beyond the borders of Colorado and contains higher rank coals as well as lignite. A deposit of lignite within it is called the Denver lignite zone [2]. Within the lignite zone is the lignite E bed, also known as the Watkins bed. It is also sometimes called the Watkins field or the Watkins lignite area. At 54.5 feet this is the thickest of the lignite deposits in the Denver basin. There is also the Wolf bed (28 feet thick) and a number of thinner lignite beds. Reference [2] gives the following ranges for properties of lignites from the Denver basin: calorific value 4000–7000 BTU lb^{-1} (9–16 MJ kg^{-1}); ash 8–30%; sulphur 0.2–0.6%.

At the Leyden lignite mine in Colorado there was a fire in 1910 which caused ten deaths [3]. The mine, owned by the Leyden Lignite Company, remained productive until 1959. At cessation of activity 4 million cubic metres of space had been created by coal removal, and there were proposals to use this for gas storage [4].

Moving to row 2 of the table, Morwell coal (see section 10.2) town gas was made from brown coal by the Lurgi process [5], which involves treatment with steam and water under high pressure (≈ 30 bar). In this process steam and oxygen are admitted to a fixed bed of coal, and the following reactions occur:

$$C + O_2 \rightarrow CO_2$$

$$CO_2 + C \rightarrow 2CO$$

$$H_2O + C \rightarrow CO + H_2$$

There was also a substantial bonus by way of some methane which arose by one or both of the following mechanisms: coal devolatilisation and tar cracking. At the beginning of its operation the Lurgi plant in Victoria produced 15 million cubic feet (0.4 million cubic metres) per day and by this means serviced the population of a city which at that time had a population of a little under 2 million. Having been piped to Melbourne the gas was blended with gas from the previously existing gas works which had used black coal brought from New South Wales, and this practice continued. The calorific value of the gas as supplied was 500 BTU $ft^{-3} \equiv 19$ MJ m^{-3}. The Lurgi gasifier in its first year of operation produced 2.2 million gallons of liquid products [6]. In the fiscal year 1960–61, 123 709 tons (≈ 11 000 tonnes) of briquettes were gasified at the Morwell facility [7]. Some musings on these figures are given in the box.

Taking the gas to be composed of carbon monoxide, hydrogen and methane and noting that the calorific values of these are[23] respectively 11, 11 and 37.5 MJ m^{-3}:

$37.5\,x + 11(1 - x) = 19$

where x is the proportion of methane

↓

$x = 0.30$

The composition of the gas is then 35% H_2, 35% CO and 30% methane.

Molar equivalence of CO and H_2 has been assumed on the basis of the chemical equation immediately above, a point addressed at the end of the calculation.

23 The calorific values on a volume basis of carbon monoxide and hydrogen differ by 1% so can be treated as equal in fuel technology calculations. This is not of course so if they are expressed on a weight basis.

The molar mass of the product gas is:

$[(0.35 \times 2) + (0.35 \times 28) + (0.3 \times 16)]$ g = 15.3 g.

This is close to the molar mass of the least abundant constituent, methane.

1 cubic metre of any gas or gas mixture at 25°C, 1 bar pressure contains 40 moles. The weight of the initial annual production of the gasification plant was then:

0.4×10^6 m^3 day^{-1} $\times$ 365 day year^{-1} $\times$ 40 mol m^{-3} $\times$ 0.0153 kg mol^{-1} $\times 10^{-3}$ tonne kg^{-1}

= 90 000 tonnes.

Addressing the 2.2 million (Imperial) gallons of liquid products referred to, this converts to 10 000 cubic metres and, at a density of about 900 kg m^{-3}, a mass of 9000 tonnes or 10% of the amount of gas. This is an intuitively reasonable result.

Adding the masses together gives 99 000 tonnes, which can be compared with the 11 000 tonnes of briquettes used in 1960–61, four to five years into the operation of the plant. So the figures fit together semi-quantitatively.

A more precise approach would involve examination of the hypothesis that the CO and H_2 are equimolar in the product gas. Amounts of these from devolatilisation would be expected to be small.

It was intended that the gasification plant at Morwell would be in service for much longer than it actually was. This of course was due to the availability of natural gas and applies also to Germany, where once there was major town gas production from brown coal. Of much more importance in the 21st century is synthesis gas from lignite and discussion of this follows.

13.3 Synthesis gas

By 'synthesis gas' (a.k.a. syngas) is meant conversion of a liquid (e.g. heavy fuel oil) or solid fuel to gas as a route to chemicals production. The basic reaction is of carbon with steam:

$$\mathbf{H_2O + C \rightarrow CO + H_2}$$

To produce say methanol from the gas the proportions of carbon monoxide and hydrogen would need to be adjusted, so that methanol could be formed according to:

$$\mathbf{CO + 2H_2 \rightarrow CH_3OH}$$

For a particular proposed synthesis therefore the synthesis gas composition needs to be controlled. Anticipating a later chapter, for Fischer-Tropsch (F-T) synthesis a H_2:CO ratio of about 2 is required. F-T requires synthesis gas as reagent, and syngas production is the theme of this section. F-T *per se* features in a later chapter as does the other route to liquid fuel production, which is hydrogenation. The scope of chemical production from syngas has extended and there is significant current use of lignites in this way. Table 13.2 gives examples, some of which are up-and-running plants and some of which are smaller scale investigations.

Table 13.2 Synthesis gas from lignites.

LOCATION AND REFERENCE	DETAILS
Indonesia [8]	Char from lignite converted to syngas with steam in the presence of a potassium carbonate catalyst. Laboratory scale. The highest H_2:CO ratio 1.682
Schwarze Pumpe site, Germany [9]	Syngas made from waste mixed with lignite used to manufacture methanol
Mongolia [10]	Experimental-scale steam gasification of lignite from Baganuur in original and ion-exchanged form
North Dakota [11]	Ammonia from synthesis gas made from lignite
Victoria, Australia [14]	Development proposals for urea production from synthesis gas made from brown coal. Target production 0.5 million tonnes per annum. Urea production involves the reaction of ammonia obtained from synthesis gas with carbon dioxide

In [8] pyrolysis of the lignite was at 850°C, and referring back to section 12.3.2 it is of interest to note that its surface area (BET) was 168 $m^2\ g^{-1}$. The point made above that F-T requires an H_2:CO ratio of about 2 is also made in [8], and the value most closely approaching 2 was achieved at a steam/carbon molar ratio of 2.2. The yield of this gas was 0.353 mol per g char, and a mass balance is attempted in the box.

A gas of H_2:CO ratio 1.682 has proportions 0.627 of H_2 and 0.373 of CO.

The molar mass is then: $[(0.373 \times 28) + (0.627 \times 2)]$ g = 11.7 g.

0.353 mol of the gas has a mass of 4.1 g and was obtained from 1 g char which, approximated to pure carbon, is 1/12 mol = 0.083 mol.

Moles of steam = $2.2 \times 0.083 = 0.183$ mol or 3.3 g.

Total mass of reactants = 4.3 g.

Mass of product = 4.1 g.

The difference of 5% can be attributed to side reactions.

At Schwarze Pumpe (second row) municipal and industrial wastes mixed with lignite from Lusatia are gasified in oxygen. Conversion of the syngas so produced to methanol is in a catalytic reactor. The annual quantity of methanol yielded at the plant is 120 000 tonnes. In the work on Baganuur lignite (next row) the minerals removed on acid treatment were believed to have promoted H_2 and CO_2 production and to have inhibited CO production. Scope for syngas composition control by varying the amounts of minerals in the feedstock lignite is indicated. Ammonia production from North Dakota lignite via synthesis gas (next row) uses coal from the Freedom mine to make synthesis gas. The reaction:

$$CO + H_2O \rightarrow CO_2 + H_2$$

takes place, and this is followed by carbon dioxide removal. The hydrogen is reacted in a suitable proportion with nitrogen to form ammonia, and this is of course well known. Modern features of the North Dakota plant include synthesis gas cleaned by the Rectisol® process [12]. The plant produces 400 000 tonnes per year of ammonia and can be viewed in [10].

One of the major current uses of ammonia is as a refrigerant, and that it should be supplied in an anhydrous state is important for that. Ammonia from synthesis gas is expected to be anhydrous. Even so, in considering ammonia from coal gasification it has to be remembered that pyrolysis is also, correctly, referred to as partial gasification, so any ammonia formed by that route needs to be considered. One report from nearly a century ago addresses this point in relation to lignite from Louisiana [13]. It is reported that under proximate analysis conditions 'aqueous ammonia' was yielded at 20–30%.

A digression into Baganuur lignite, which features in Table 13.2, follows. It was pointed out in a number of earlier sections including 10.3 how uncertainties in classification between lignite and sub-bituminous coal can arise. In [15] this matter was addressed in relation to Baganuur where lignite and sub-bituminous coal co-exist. Different samples from Baganuur having widely varying carbon contents (from 66.7 to 84.3%) and oxygen contents (from 11.6 to 27.4%) were examined using carbon-13 n.m.r. By comparison of the spectra for samples high in carbon with those from samples lower in carbon and observation of changes in skeletal carbon structure, conclusions on the mechanism of oxygen loss in going from lignite to sub-bituminous could be drawn. The mechanism involved loss of methoxy phenols and of dihydric phenols. Model compounds representing these would be respectively mequinol (4-methoxy phenol) and pyrocatechol (*o*-dihydroxy benzene).

13.4 Synthetic natural gas

The Dakota Gasification Company produces daily 145 million cubic feet (4 million cubic metres) of synthetic natural gas (SNG) with lignite as feedstock [16]. Some background on the production of synthetic natural gas is needed.

First, the carbon content of the lignite is gasified with steam according to:

$$C + H_2O \rightarrow CO + H_2$$

Reaction of hydrogen with the CO gives:

$$CO + 4H_2 \rightarrow CH_4 + 2H_2O$$

There are variants on the above chemical equations, including invocation of a reaction between carbon dioxide and hydrogen, but such details are not seen as important in a discussion such as this. More important is the source of the hydrogen. Where does it come from? It can in fact easily be made by steam reforming of low-value refinery products. Such a product will have an elemental composition approximating to CH_2, and hydrogen can be made from it by:

$$\text{'}CH_2\text{'} + H_2O \rightarrow CO + 2H_2$$

That of course is not the only way of getting elemental hydrogen for chemical processing: electrolysis is another. Other examples of SNG production from lignites are given in Table 13.3.

Table 13.3 Examples of SNG production by gasification of lignites.

LOCATION	DETAILS
Beulah, North Dakota[24] [17]	14 500 tonnes per day of lignite converted to SNG
Eastern Mongolia [18]	Plant under construction which will produce 4 billion cubic metres annually of SNG from locally mined lignite
Kemper County, Mississippi [19]	Conversion of local lignite to SNG

In the introduction to this section it was mentioned that carbon dioxide sometimes features in the process, and this is so at Beulah where carbon dioxide is reacted with hydrogen over a catalyst to form methane ('methanation'). However, the process is set up so that there is, so to speak, carbon dioxide to spare, and this is pipelined across the border to Canada for enhanced oil recovery. Use of pre-SI units is still prevalent in the USA, and we are told in [17] that the SNG produced from the Beulah lignite has calorific value 975 BTU per cubic foot, which converts to 36.9 MJ m^{-3}. The plant in Beulah pre-dates that operated by the Dakota Gas Company and was in fact the first such plant in the world. It began production in 1984.

Moving on to the next row, it should be noted that at the plant in eastern Mongolia there will be a tar by-product. This is not uncommon in gasification of low-rank coals and is recorded above to have occurred at the Morwell gasification plant in Australia where the desired product was town gas. The next row is concerned with Mississippi, where the SNG produced will be blended with conventional natural gas in power generation.

The pie chart in Figure 13.1 originates in North Dakota [20] and shows the breakdown of lignite usage there. Electricity generation accounts for over three quarters. Fertiliser

24 See also section 6.2.

(ammonia) and SNG production each require synthesis gas. The pie chart links this section to section 6.2.

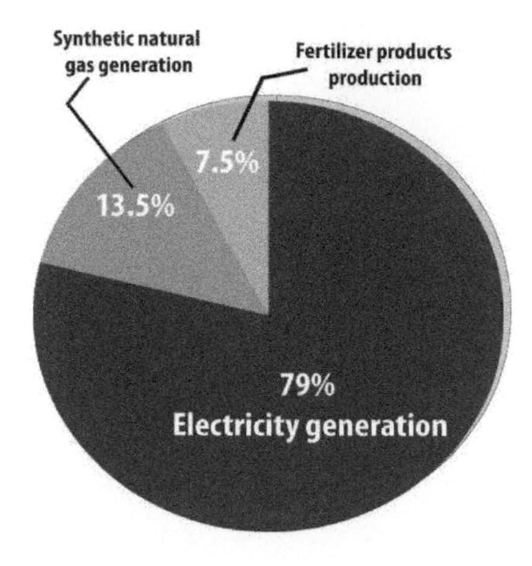

Figure 13.1 Pie chart showing breakdown of lignite usage. Taken from [20].

13.5 Lignite char gasification

In considering lignite char separately from lignite, one has to be aware that carbonisation is part of lignite gasification and either precedes it or occurs simultaneously with it. Even so it is of interest to consider gasification with starting material already carbonised. A carefully selected and fairly recent example of such a study will be drawn on here [21]. Three lignite chars and three chars from bituminous coals[25] were examined for reactivity in gasification with steam by means of a fluidised bed and a thermobalance. The desired reaction was simply:

$$C + H_2O \rightarrow CO + H_2$$

and this chemical equation recurs in this chapter. The authors of [21] are quite unequivocal in asserting the higher reactivity of the lignite char vis-à-vis the bituminous coal coke. This is a general trend: the difficulty of using lignite char as a metallurgical reductant because of its high reactivity was noted in section 12.4.

13.6 Producer gas from lignite

Producer gas came into use in the 19th century. It is made by passing air or air/steam through a bed of coal or coke. Because air is used the fuel gas resulting is heavily diluted with nitrogen, and its calorific value is around 5 MJ m^{-3}. Over the period 1870–1940 producer gas became the mainstay of certain industries including glass manufacture,

25 Reference [21] describes those from the bituminous coals as chars, but according to the nomenclature in section 12.1 of this text they would be better described as 'cokes' or perhaps as 'coke particles'.

but it had diminished in importance by about 1950 by which time natural gas and LPG had become available. There has been a resurgence of interest in recent years, and R&D has focused on producer gas from biomass starting material so as to produce a gas from which carbon credits can be generated.

Articles on producer gas in the present-day literature are few and far between. It is probably better to examine a report from the hey-day of producer gas for discussion, and the one chosen is [22] in which lignites from Texas, North Dakota and Arkansas feature. Here it is reported that ND lignite proved, at a United States Geological Survey test facility, to be a good feedstock for producer gas, and its adoption by electricity producers is advocated. Usually when producer gas is used in power plant a gas turbine rather than a steam turbine is in operation. This is so with the modern applications of producer gas, previously referred to, where carbon mitigation is required [23].

13.7 Gasification with supercritical water

In very recently reported work [24] hydrogen production by gasification of a lignite from Inner Mongolia is described. Gasification was with steam as 560°C, 25 MPa, conditions which approach ultra-supercritical according to the definition given in the Appendix to Chapter 4. Hydrogen was formed by the steam–carbon reaction and by the water gas shift reaction. With a suitable reactor residence time it was possible to obtain gas of up to 65.6% molar basis hydrogen, balance carbon dioxide.

13.8 Co-gasification

Co-gasification of lignite with other substances including plastic waste and household waste takes place at the Schwarze Pumpe methanol plant in Germany. It produces 140 000 tonnes per year of methanol [25] by passing the synthesis gas so made over a catalyst. Wood waste is sometimes co-fired with the lignite. This provides for methanol to some degree 'carbon-neutral by paternity' and one would expect that to be a plus in the price received for the methanol. At Berrenrath in Germany there has been, on a demonstration basis, co-gasification of local lignite with household waste [26] again to make methanol. It was concluded that gasification of a 50:50 lignite–household waste blend was viable. Here again there could be a bonus in terms of carbon neutrality as some of the components of household waste including paper and cardboard are themselves carbon-neutral.

There has been interest in Greece in co-firing lignite with waste materials in the production of electricity [27]. The waste proposed for use is refuse-derived fuel (RDF), household waste having been treated by drying, shredding and pelletising.

Reference [28] describes laboratory-scale co-gasification of a lignite and a petroleum coke at 1000°C, as well as gasification of the two separately. Heating to that temperature was under nitrogen, whereupon the atmosphere was changed to carbon dioxide. In some tests the solid mixture was dry at the commencement of heating, there being gasification by pyrolysis only up to 1000°C. In other tests the solid was in the form of a slurry in water, so pyrolysis was accompanied by the reaction:

$$C + H_2O \rightarrow CO + H_2$$

Once the temperature was at 1000°C it was held there for 100 minutes and the nitrogen was changed for carbon dioxide, enabling the reaction:

$$C + CO_2 \rightarrow 2CO$$

to occur. At time 100 minutes the petroleum coke only had lost only about 20% of its weight. Petroleum coke and lignite in a dry blend showed about a 60% weight loss. Petroleum coke and lignite as a slurry showed about 90% weight loss whilst the lignite lost its entire organic content after one hour, only ash remaining. A trend can be discerned in these results. The petroleum coke is capable only of very limited pyrolysis, and the reaction with carbon dioxide falls a long way short of being complete. Lignite only is reactive towards pyrolysis and its char towards reaction with carbon dioxide. Blends of the two show the expected intermediate behaviour.

The point was made in section 13.3 that pyrolysis – the basis of carbonisation – is also partial gasification. In [29] a Thai lignite is pyrolysed with corncob, and the product of primary interest is gas, so this is also describable as co-gasification. The investigation, on a laboratory scale with a thermogravimetric analyser, co-pyrolysed a lignite from Thailand with corncob in ratios from 90% lignite to 10% lignite at temperatures up to 600°C. As would be expected total gas yield was greater for corncob alone than for lignite alone, a 50:50 blend coming in between. Gas components which will make for a fuel gas are carbon monoxide, hydrogen and methane. Methane is particularly desirable as it has a calorific value on a molar basis three times that of the other two. At 400°C the 50:50 blend showed a surprisingly high yield of methane, higher than for either the lignite or the corncob alone. Even so methane yields were about an order of magnitude lower than carbon monoxide yields.

Another example of lignite–biomass gasification was of wood pellets with Rhenish lignite in a fluidised-bed gasifier [30]. Gasification temperature was 850°C, and coal:wood mass ratios were across a range. In going from 100% wood to 100% lignite in the feedstock, the hydrogen content of the gas increased linearly from just under 35% to $\approx$ 50%. The carbon monoxide decreased from 35% to $\approx$ 30%, whilst methane decreased from 10% to just over 5%. The highest calorific value would have been for the gas from wood pellets only because of the high methane, and there is the bonus that this methane, being derived from biomass, is carbon-neutral.

Co-gasification in a fluidised bed of Loy Yang brown coal with pelletised biomass from an algae species is reported in [31]. Both macroalgae and microalgae were so used at 10% of the coal weight and comparisons were made with results from gasifying the coal only. With the macroalgae pellets there were rises in the hydrogen and carbon monoxide contents of the product gas, with lowering of the carbon dioxide. The microalgae pellets had the effect of lowering the total yield of gas, an effect clearly due to the remarkably high ash content (37.8%) of the pellets from the microalgae used which was of the genus *Scenedesmus*. With neither type of biomass pellet was the methane content affected.

13.9 Underground gasification of lignite

13.9.1 Introduction

This section will first examine a scene of underground gasification of lignite which has been active for over half a century. More modern projects into such gasification in various parts of the world will then be discussed.

13.9.2 Angren, Uzbekistan

Reference to this activity was made in section 9.4. The underground gasification of coal (UGC) plant at Angren is one of only two remaining in the FSU. Previously there were a large number, the products of which have been replaced by natural gas. The plant at Angren, which uses lignite, came into operation in 1961 [32]. Gasification is with air, so the quality of the gas is no higher than that of producer gas. This is confirmed in [33] where the calorific value of the gas is given as 800–1000 kcal $m^{-3} \equiv 3.4$–4.2 MJ m^{-3}. The point is made even more strongly when it is stated that the gas is about 50% nitrogen. This does not of course preclude use of the gas in electricity generation; this follows from what was said about producer gas in section 13.6. However, in [34] it is stated that heat from the gas produced at Angren costs 3 roubles per Gcal. Calculations are given in the box.

> Noting that [33] was published in 2013, we compare the cost given with that of the average Henry Hub price for natural gas for the middle of that year, which was $US3.7 per million BTU of heat [35]. We also note that 10^6 BTU $\approx$ 1 GJ. The price therefore is $US3.7 per GJ.
>
> Using a factor of 0.02 for conversion of the rouble to the US dollar, the cost given in [33] for heat from the manufactured gas converts to $US0.25 per GJ.

It is clear that purely on a heat basis the gas from Angren compares well. Use of Henry Hub as a benchmark is reasonable, as at any one time Russian natural gas will be selling on world markets for a price not widely different from Henry Hub. However, there are other factors to be considered. One is that unit heat from natural gas costs about a fifth of unit heat from a distillate from crude oil [36], and that is why so much natural gas used to be flared before greenhouse gas mitigation placed restrictions on the practice. Notwithstanding the factor of 13 between the Henry Hub price and that of Angren gas on a heat basis, the author of [34] declares that Angren gas is not competitive with natural gas.

13.9.3 Current projects in underground gasification of lignite

In Turkey lignite from the Thrace basin is being evaluated for underground gasification. Chemical and petrographic details of two lignites from Thrace are given in [37]. Each has an ash content in excess of 20% and a sulphur content of 1–2%. In both, huminite at about 70% is by far the dominant maceral group. The primary activity at Thrace is natural gas production, and the lignite there was discovered whilst drilling for gas [38].

Plans for underground gasification of brown coal in Victoria, Australia are clearly focused on synthesis gas rather than fuel gas [39]. This is possibly a factor relevant more widely than just to Victoria: the product of underground gasification of coal is likely to be more saleable as a chemical feedstock than as a gaseous fuel. It therefore competes with feedstocks from petroleum. In Victoria methanol production, ammonia production and F-T are all seen as uses to which the syngas could be put. In South Australia, Leigh Creek coal has been considered for underground gasification. There was a study into it as far back as 1983, and revived interest in the early 21st century, for example [40] where results of the 1983 study were invoked in making a case that gas from this source would be cheap. Experience in the FSU was similarly drawn on.

In North Dakota, Harmon lignite from the west of the state has recently been evaluated for underground gasification [41]. The emphasis is less on the nature of the coal than on the identification of suitable sites in the deposit for the gasification process. Relevant to this is the point expressed in [42] that the cavity within the coal deposit at which gasification takes place functions as a chemical reactor. In Mongolia there is underground gasification of lignite at the Dayan deposit (see section 15.3.2). In the part of the former Yugoslavia now called Slovenia there was once interest in underground gasification of the lignite at Velenje (see section 5.8.3), and an investment study was carried out in 1960 [43]. The idea faltered because of competition from oil and natural gas, and in 1968 a government decision was taken not to continue.

13.10 Coal-bed methane from lignites

13.10.1 Introduction

Coal-bed methane (CBM) is one of the great facts of our era, major activity in it having coincided with the introduction of tight gas from inorganic geological structures. There is a major difference between CBM and gas from an inorganic formation whether tight or conventional: the former contains no condensate, and this can be a factor in the viability of its collection. This section will be concerned with CBM from lignite deposits.

13.10.2 Examples

Table 13.4 gives details of CBM in lignite deposits in various parts of the world. The table also has one entry of the production of methane from lignite by bacterial action.

Table 13.4 Coal-bed methane in lignite deposits.

LOCATION AND REFERENCE	DETAILS
Louisiana, USA [44]	Exploratory drilling for CBM at scenes including Dolet Hills
North Dakota, USA [45,46]	No CBM production at the present time. Large quantities known to exist in the Williston basin
Thar field, Pakistan [47]	Experimental studies with methanogenic bacteria. See main text
Kazakhstan [48]	75–100 billion cubic metres of methane believed to exist at Ekibastuz (see section 9.3)
South Australia [50]	Leigh Creek coal examined for CBM potential with positive results
Sumatra, Indonesia [52]	Biogenic CBM from six newly drilled wells in a lignite deposit

The investigation in Louisiana involved desorption analysis of methane from lignite samples, for which a figure of 11 standard cubic feet of methane per ton of dry, ash-free lignite was determined. This converts to 0.34 cubic metres of methane per tonne of lignite. The Williston basin in ND (next row) contains both lignite and sub-bituminous coal in an estimated total quantity of 530 billion tons. The CBM obtainable from it is estimated as 13 Tft^3. This converts to 0.76 cubic metres of methane per tonne of coal. A lignite field will contain pockets of methane which will, on a solubility model of coal–gas interaction, be in phase equilibrium with methane in the lignite structure. That is one reason why the figures for desorbed gas and total gas differ, with the latter obviously being higher.

The term 'methanogenic bacteria' is fairly self-explanatory: these microorganisms produce methane in the course of their metabolism of an organic medium, and require anaerobic conditions to do so. When tested with methanogenic bacteria, lignite from the Thar deposit yielded methane in the range 2.13–16.33 standard cubic feet per ton of coal equivalent to 0.07–0.5 cubic metres per tonne of coal. These are of the same order as the range of the figures given in the previous section, though it must be remembered the two sets of numbers relate to different phenomena so the basis for comparison is limited. CBM can be either biogenic or thermogenic in origin. Thermogenic CBM occurs at greater depths into a deposit than biogenic.

Any CBM from Kazakhstan (following row) could obviously be exported by means of the Central Asia to China gas pipeline [49], adding to the other gas conveyed by that facility which has a capacity of 55 billion cubic metres per annum. In fact the prospects of particular sources of CBM depend on proximity to pipelines installed for conventional natural gas and their ability to take extra capacity. Moving on to the Leigh Creek coal, the question is raised in [50] of the extent to which the high moisture content of the coal will act against methane containment, and a start in addressing this is the fact that bituminous coals from Kentucky were found to yield 1.3–1.7 cubic metres of CBM per tonne [51], higher than the values given for Louisiana lignite but not by anything like an order of magnitude. A wider and deeper survey of CBM yields for coals of different rank would need to be on a multivariate basis, taking in water content as well as quantities relevant to the solubility approach briefly referred to above. These include the solubility parameter of a particular coal and how close it is to that of methane. In reference [52] in the final row there is mention of the practice of nutrient injection into a well to promote biogenic methane formation.

A North Dakota lignite was examined for methane release in work described in [52], the purpose of which was an investigation of the mechanism of CBM formation. Samples of the coal were pyrolysed across a range of temperatures and the methane yield measured. At the highest pyrolysis temperature of 727 K, the production of methane scaled from laboratory amounts to industrial was 1560 standard cubic feet per ton (48 m^3 per tonne). This of course exceeds the values given previously in this section because of the high temperature. The authors of [52] point out that this figure can be scaled to bed temperatures from chemical kinetic information.

In [53] changes in the coal as a result of methane production were also monitored, there being an awareness of the high atomic ratio of hydrogen (4:1) in methane. Accordingly methane loss in the experiments was accompanied by increased apparent maturity manifest as huminite reflectance, a quite dramatic rise from an initial value of 0.31 to 1.61 for the coal having experienced 727 K.

13.11 Concluding remarks

The next chapter, on liquid fuel production from lignite, follows naturally from this one. The point has been made that F-T, a major route to liquids from coal, requires synthesis gas as reactant. The overlap of content is therefore clear.

13.12 References

[1] http://www.boulderweekly.com/article-11230-boulderrss-town-gas-processes-and-the-resulting-contamination.html

[2] Nichols D.J. *Summary of the Tertiary Coal Resources of the Denver Basin, Colorado*, Professional Paper 1625-A, United States Geological Survey (1999)

[3] http://historicleyden.org/?q=node/9

[4] Meddles R.M. 'Underground gas storage in the Leyden lignite mine', *1978 Symposium Rocky Mountain Association of Geologists* pp. 51ff.

[5] http://www.britannica.com/EBchecked/topic/122944/coal-utilization/81733/The-Lurgi-system

[6] https://news.google.com/newspapers?nid=1300&dat=19561206&id=jlBVAAAAIBAJ&sjid=dpUDAAAAIBAJ&pg=6348,817523&hl=en

[7] *Eleventh Annual Report*, Gas and Fuel Corporation of Victoria, Melbourne, Vic. (1961)

[8] Supramonoa D., Tristantinia D., Rahayua A., Suwignjoa R.K., Chendraa D.H. 'Syngas production from lignite coal using K_2CO_3 catalytic steam gasification with controlled heating rate in pyrolysis step', *Procedia Chemistry* 9: 202–209 (2014)

[9] McPhail S.J., Cigolotti V., Moreno A. *Fuel Cells in the Waste-to-Energy Chain: Distributed Generation Through Non-Conventional Fuels and Fuel Cells*, Springer (2012)

[10] http://rlhxxb.sxicc.ac.cn/EN/abstract/abstract18160.shtml#

[11] http://www.dakotagas.com/Gasification/Ammonia_Plant/

[12] http://www.linde-engineering.com/en/process_plants/hydrogen_and_synthesis_gas_plants/gas_processing/rectisol_wash/index.html

[13] Glenk R. *Louisiana Lignite*, Bulletin No. 8, Department of Conservation, Division of Economic Geology, State of Louisiana, New Orleans (1921)

[14] http://www.latrobefertilisers.com.au/project_latrobe.html

[15] Erdenetsogt B., Lee I., Lee S.K., Ko Y., Bat-Erdene D. 'Solid-state C-13 CP/MAS NMR study of Baganuur coal, Mongolia: oxygen-loss during coalification from lignite to subbituminous rank', *International Journal of Coal Geology* 82: 37–44 (2010)

[16] http://www.dakotagas.com/Gasification/

[17] http://www.netl.doe.gov/research/coal/energy-systems/gasification/gasifipedia/great-plains

[18] http://www.chinagasholdings.com.hk/siteen/aspx/News_Infor.aspx?id=649

[19]

http://www.naturalgasintel.com/articles/99399-mississippi-power-to-process-in-state-coal-to-fuel-gas-fired-kemper-county-unit

[20] https://www.lignite.com/about-lignite/lignite-uses/

[21] Song B., Zhuy X., Moon W., Yang W. 'Steam gasification of low-rank coal chars in a thermobalance reactor and a fluidised bed reactor', *13th International Conference on Fluidization: New Paradigm in Fluidization Engineering*, Engineering Conferences International (2010)

[22] Professional Paper – United States Geological Survey, Issue 100 USGS (1917)

[23] Rabou L.P.L.M., Grift J.M., Conradie R.E., Fransen S., Verhoeff F. 'Micro gas turbine operation with biomass producer gas', *15th European Biomass Conference*, Berlin (2007)

[24] Jin H., Guo L., Guo J., Ge Z., Cao C., Lu Y. 'Study on gasification kinetics of hydrogen production from lignite in supercritical water', *International Journal of Hydrogen Energy* 40: 7523–7529 (2015)

[25] http://www.icis.com/resources/news/2006/11/08/1104202/sustec-schwarze-pumpe-restarts-methanol-plant/

[26] *Gasification Technologies*, ThyssenKrupp Uhde (n.d.)

[27] Koukouzas N., Katsiadakis A., Karlopoulos E., Kakaras E. 'Co-gasification of solid waste and lignite – a case study for Western Macedonia', *Waste Management* 28: 1263–1275 (2008)

[28] Zhan X., Jia J., Zhou Z., Wang F. 'Influence of blending methods on the co-gasification

reactivity of petroleum coke and lignite', *Energy Conversion and Management* 52: 1810–1814 (2011)

[29] Sonobea T., Worasuwannaraka N., Pipatmanomaia S. 'Synergies in co-pyrolysis of Thai lignite and corncob', *Fuel Processing Technology* 89: 1371–1378 (2008)

[30] Kern S., Pfeifer C., Hofbauer H. 'Co-gasification of wood and lignite in a dual fluidized bed gasifier', *Energy & Fuels* 27: 919–931 (2013)

[31] Zhu Y., Piotrowska P., van Eyk P.J., Boström D., Kwong C.W., Wang D., Cole A.J., de Nys R., Gentili F.G., Ashman P.J. 'Co-gasification of Australian brown coal with algae in a fluidized bed reactor', *Energy & Fuels* 29: 1686–1700 (2015)

[32] http://www.lincenergy.com/acquisitions_yerostigaz.php

[33] http://www.eioba.com/a/1ws2/underground-coal-gasification-in-republic-of-uzbekistan

[34] Simeons C. *Coal: Its Role in Tomorrow's Technology*, Elsevier (2013)

[35] http://www.eia.gov/dnav/ng/hist/rngwhhdw.htm

[36] Jones J.C., Russell N.V. *Dictionary of Energy and Fuels*, Whittles Publishing, Caithness and CRC Press, Boca Raton (2007)

[37] Toprak S. 'Petrographic properties of major coal seams in Turkey and their formation', *International Journal of Coal Geology* 78: 263–275 (2009)

[38] http://www.linknovate.com/publication/underground-coal-gasification-and-applicability-to-thrace-basin-lignite-in-turkey-2378581/

[39] http://www.energyandresources.vic.gov.au/energy/about/publications-and-resources/low-emission

[40] Walker L.K. 'The future role for underground coal gasification in Australia', *National Conference*, Australian Institute of Energy (2006)

[41] Pei P., Zeng Z., He J. 'Characterization of the Harmon lignite for underground coal gasification', *Journal of Petroleum Science Research* 3: 136–144 (2014)

[42] http://www.worldcoal.org/coal/uses-of-coal/underground-coal-gasification/

[43] Zvezi P.R.V., Premoga S.P.U., Slovenij V. 'Overview of energy studies on UCG in Slovenia', *Journal of Energy Technology* 7: 55–56 (2014)

[44] Breland F.C. *Selected Presentations on Coal-Bed Gas in the Eastern United States* (P.D. Warwick, ed.) pp. 27–35, United States Geological Survey (2004)

[45] http://www.nap.edu/openbook.php?record_id=12915&page=19

[46] http://pttc.mines.edu/casestudies/williston/index.html

[47] Haider R., Ghauri M.A., Jones E.J., San Filipo J.R. 'Methane generation potential of Thar lignite samples', *Fuel Processing Technology* 126: 309–314 (2014)

[48] CMM Country Profiles – 20 Kazakhstan

[49] http://www.cnpc.com.cn/en/FlowofnaturalgasfromCentralAsia/Flowofnaturalgasfrom-CentralAsia2.shtml

[50] Sansome A., Nitschke L., Tingate P.R. 'Coal seam methane potential of South Australia',

MESA Journal 47: 11–13 (2007)

[51] http://www.uky.edu/KGS/coal/cbm.htm

[52] http://www.naturalgasasia.com/next-fuels-china-indonesia-coal-to-gas-pilot-well-update-5031

[53] Tang Y., Jenden P.D., Nigrini A., Teerman S.C. 'Modeling early methane generation in coal', *Energy & Fuels* 10: 659–671 (1996)

CHAPTER 14
CONVERSION TO LIQUID FUELS

14.1 Introduction

This chapter will examine carefully selected accounts of the production of liquid fuels from lignites over the 70 or more years that this has been taking place. Fischer-Tropsch (F-T) and hydrogenation by the Bergius method will both be considered.[26] As recorded in section 13.3, the F-T process starts with synthesis gas which can be made from coal, from liquid hydrocarbons or from natural gas. Steps involved in going from synthesis gas to hydrocarbons include:

$$n\mathrm{CO} + (2n + 1)\,\mathrm{H_2} \rightarrow \mathrm{C}_n\mathrm{H}_{2n+2} + n\mathrm{H_2O}$$

and the reader is referred to basic coverages of the topic (e.g. [1,2]) for details of the other reactions occurring. F-T always requires a catalyst, and continuing research into it has been largely on the catalysis side.

This is a suitable place to discuss yields of oil in F-T processes involving lignites. The calculation shown in the box is very rough but gives a ball-park figure on yields expected.

> A lignite in its bed-moist state will have a carbon content in the neighbourhood of 50%. A tonne of lignite in this state therefore contains 500 kg of carbon.
> A reliable rule in stoichiometric calculations and the like is that all petroleum fractions approximate to CH_2 in empirical formula. So when a tonne of the lignite undergoes F-T the process can be summarised:
>
> 500 kg C $\rightarrow$ (14/12) $\times$ 500 kg liquid hydrocarbon = 585 kg liquid hydrocarbon.
>
> Products of F-T are less dense than those of crude oil and a value of 800 kg m^{-3} is reasonable. So the liquid yielded is in volume terms:
>
> (585 kg/800 kg m^{-3}) = 0.73 m^3
>
> Recalling that a barrel is 0.159 m^3 the above becomes 4.6 barrels.

The author has found only one web source against which to check this calculated result. Reference [3] says that in F-T, one ton (US ton) gives on F-T processing '3 to 4 barrels of oil'. So the above rough calculation errs a little on the high side.

14.2 F-T activity in Germany at the time of World War II

Over the period of interest in this section, Bergius hydrogenation was found to be more suitable as a means of making liquid fuels from German lignites than F-T. Even so F-T

26 Of course F-T is itself a hydrogenation process, but the term when applied to liquid fuel production from coal usually means application of the Bergius process, to be discussed later.

did take place on a major scale, and Table 14.1 summarises lignite conversion to liquid fuels in Germany by F-T at that time. Catalysts were iron or cobalt based.

Table 14.1 Scenes of liquid fuels production from lignites by F-T in Germany.

LOCATION	DETAILS
Leuna, Lower Saxony	Commencement of operations in 1927. Production over the first year 30 000 tons (27 200 tonnes) of liquid fuel [4]. (See also the note on Leuna in the opening paragraph of section 14.3)
Lützkendorf, near Mücheln	29 320 tonnes of liquid fuel product in 1944 [5]
Schwarzheide, Lusatia	164 600 tonnes in 1942 [7]
Bohlen, close to Lippendorf	See comments in the main text

The quantity of product over the first year given in the table converts to ≈ 200 000 barrels. The Lützkendorf–Mücheln facility (following row) sourced lignite from the nearby Lützkendorf pit. There was a catalyst production facility there, which produced cobalt-based catalysts for other F-T plants [6]. The facility at Schwarzheide was operated by Brabag[27] [8], formed when a group of lignite producers, under duress, financed the production of liquid fuels from lignite in readiness for war.

It was described in section 13.2 how the Lurgi process in Victoria yielded significant amounts of liquid products. This is also sometimes true of F-T, especially with high-volatile coals which of course lignites are almost by definition. At Bohlen (row 4 of the table) tars resulting in this way were hydrogenated to make motor fuels [9].

14.3 Bergius hydrogenation activity in Germany at the time of World War II

The Bergius process, the centenary of which was in 2013, can be summarised [10]:

$$n\mathrm{C} + (n + 1)\,\mathrm{H_2} \rightarrow \mathrm{C}_n\mathrm{H}_{2n+2}$$

and conditions of high temperature (400–500°C) and pressure (up to 700 bar) are involved as well as a catalyst. Jointly with C. Bosch, Bergius received the 1931 Nobel Prize in Chemistry [11]. Another collaborator was Matthias Pier (1882–1965), and the Bergius process is sometimes referred to as the Bergius–Pier process. In reviewing the information in the previous section and in this one, a reader should be aware that at some industrial centres in Germany at the period of interest both F-T and Bergius processes, each using lignite, were taking place. This is actually true of Leuna and of Bohlen [12]. Plants in Germany at that time which produced liquid fuels by hydrogenation of lignite or of lignite tars, obtained from carbonisation, include that at Magdeberg, operated by Brabag, where lignite tar from low-temperature carbonisation to make 'granular coke'[28] was hydrogenated by the Bergius process [13]. The capacity of the

27 Braunkohlen Benzin AG.
28 See section 12.3.2 for clarification of terminology.

plant was 230 000 tons (≈ 1.5 million barrels) per year. There was also a plant at Wesseling in Rheinland which, using previously unprocessed lignite, produced a comparable quantity [14]. In 'whole coal' hydrogenation the petrographic composition is relevant; vitirine/huminite promotes hydrogenation [15]. In concluding this section we note that production of liquid fuels from lignites in Europe did not stop with 'cessation of hostilities'. For example, during the German occupation of Czechoslovakia a brown-coal-to-liquid plant was set up there, and this was producing until the 1960s.

There is considerable current interest in making liquid fuels from lignite tar, exemplified by a recent contribution to the literature originating in the Czech Republic [16]. This work used several lignites from the Czech Republic including more than one from Sokolov (see section 5.1.4). The Czech lignites were pyrolysed, producing under the conditions used tar in yields between 7 and 19% of the original coal weight. These were processed in hydrogen gas at vessel admittance pressures up to 3 MPa in the presence of a cobalt–molybdenum catalyst manufactured by BASF. Emphasis was on analysis of organic compounds in the hydrogenated material, and it was recorded that organic compounds corresponding to those in the naphtha fraction from crude oil distillation had been identified.

14.4 More recent F-T activity

This includes an enterpise at Kazakhstan currently on the demonstration scale [17]. At the time of going to press the plant produces 0.8 tonnes per day of liquid fuel from the lignite deposit near Yerementau in central Kazakhstan. Kazakhstan is a major oil producer, in contrast to New Zealand where there is interest in liquid fuels from lignites from Otago in the South Island via F-T [18]. There is also investigative activity in Australia's Latrobe Valley. There has also been interest in liquid fuels from lignite via F-T in Mongolia where Shenhua have been conducting trials.

All of the above are however at most at the development stage, sometimes only the planning stage. In fact there is only one F-T coal-to-liquids facility in the world which operates on a full-scale, commercial basis. That is the SASOL plant at Secunda in the Transvaal, South Africa, which uses sub-bituminous coal. The reason for the lack of interest elsewhere is that gas-to-liquids processes, starting with natural gas, are strongly preferred at present.

14.5 Hydrogenation by donor solvents

14.5.1 Introduction

About 40 years ago the Exxon Donor Solvent Coal (EDS) liquefaction process came into being and its early findings are reported in [19]. It is applicable to all ranks of coal. The author of [19] makes the statement:

> *That the Bergius process was a technological success is unquestionable; whether it was an economic success is debatable.*

In the EDS process the donor solvent is generated *in situ* by hydrogenation of a petroleum-derived recycle solvent [20]. The procedure does then involve elemental hydrogen, but under milder conditions than in the Bergius process. Since then R&D into coal-to-liquids by this approach has abounded, and lignites have had an important place in this. The authors of [21] state that the role of the donor solvent is in 'shuttling hydrogen from the gas phase to the coal'. A catalyst can also have a shuttling role.

14.5.2 Summary of selected investigations

Table 14.2 gives examples of research into lignite hydrogenation by donor solvents. Obviously selection of work for inclusion has had to be arbitrary. It is at least intended that all of the regions of the world engaged in such activity will be represented and that there will be links to scenes of lignite activity described earlier in the book.

Tetralin (1,2,3,4-tetrahydronapththalene) is widely used in coal hydroliquefaction. In the work in row 1, products were distributed between pre-asphaltenes, asphaltenes and 'oil and gas'. The distinction between pre-asphaltenes, asphaltenes and oil is on the basis of solubility in certain solvents. For example, a pre-asphaltene is insoluble in benzene: an asphaltene is soluble in benzene and insoluble in *n*-pentane. Asphaltenes are of course present in crude oil.

In the work from the FSU in the second row, there is some emphasis on the effectiveness of the respective catalysts. These included haematite and magnetite, and were all obtained from iron ore and therefore inexpensive. In the work on the Chinese lignite in the following row catalyst performance was, again, of major interest. High yields in the gas and oil category of product were favoured by a catalyst composed of iron II sulphide with some added sulphur.

Petrographic factors are referred to in rows 4 and 5. Reference [26] goes on to say that no such correlation is apparent for Texas lignites treated under the same conditions, and the tentative conclusion is drawn that for them macerals other than those in the vitrinite group are involved in hydrogenation reactions. The same source states that for many US lignites ammonium molybdate is the most suitable catalyst. An investigation of hydrogenation of a higher rank coal with ammonium molybdate as a catalyst [27] used hydrogen gas only, and it is stated that there was no solvent as a 'vehicle'. This is of course another way of expressing what was described above as the shuttling role of a solvent. Clearly in [27] this was fulfilled by the catalyst alone.

Findings on a Canadian lignite, from the same source, follow in row 6. Reference [26] states that increasingly strong reaction conditions – up to 30 MPa with tetralin and a catalyst – will enable all of the petrographic components to be converted with the exception of fusain and opaque attritus. We are informed in [28] that the fusain macerals belong to the inertinite group as does opaque attritus, and there is only minor variation on the terminology in reference [26] of Chapter 2 which, thus far in the book, has been the source drawn on for petrographic nomenclature. These macerals then are too unreactive to respond even to the most vigorous hydrogenation conditions applied to the US lignites within the scope of the discussion in [26].

Table 14.2 Investigations of hydrogenation of lignite by donor solvents.

ORIGIN OF THE LIGNITE AND REFERENCE	DETAILS
Turkey [22]	Two lignites, one from Tuncbilek (see Table 1.1), treated in tetralin at temperatures in the range 325–425°C under nitrogen at 50 bar. One-litre reactor. Iron, molybdenum and iron oxide catalysts. Conversions in the approximate range 60–70%, depending on the temperature, catalyst and reacting time
FSU [23]	Kansk-Achinsk lignite (see section 9.2 et seq.) treated with hydrogen gas at 12 MPa and a solvent at ≈ 400°C. Conversions in the range 20–90%. Several catalysts tested
Yunnan Province, China [24]	Treatment with tetralin and hydrogen gas (5 MPa) at 400°C. Various catalysts. Reacting times in the range 30–60 minutes. Conversions up to 90%, resolved into pre-asphaltenes, asphaltenes and 'oil and gas'
Spain [25]	Treatment with tetralin and hydrogen of four lignites. Temperature 400°C. A 60 cm3 vessel ('microreactor'). Good conversion rates promoted by high vitrinite content. Distribution of products across the categories not affected by petrographic composition
North Dakota [26]	Hydrogenation with tetralin and hydrogen at 5.5 MPa at 430°C of a set of lignites. Linear correlation between the amounts of soluble products and the huminite content of the parent coal
Estevan mine, Saskatchewan [26]	57% conversion with tetralin at 3 × the lignite weight at 400°C
Thar field, Pakistan [29]	Evaluation of this newly developed lignite resource (see section 8.3) for hydrogenation potential. Factors noted include the following: Huminite, liptinite and inertinite contents averaged across 32 samples 88.7%, 9.0% and 2.3%. Huminite reflectance 0.33–0.4%. Within the huminite group, gelinite and ulminite the most reactive towards hydrogenation
South Banko and Kalimantan, Indonesia [30]	Five brown coals reacted with hydrogen and a donor solvent in the presence of a nickel–molybdenum catalyst and added sulphur
Banko, Indonesia [31]	Use of amines as additives
Morwell, Victoria, Australia [32]	Hydrogenation tests on lithotypes
Loy Yang, Victoria, Australia [33]	See comments in the main text
Kolubara, Serbia (see sections 5.8.4 and 24.1) [34]	Good conversion. Product high in substances soluble in n-heptane

In the evaluation of Thar lignite in the following row, emphasis is very much on the petrographic analysis. One of the points under consideration in the work on Indonesian brown coals in the following row was added sulphur and whether it enhanced conversion. This point is touched on earlier in this section in discussion of the Chinese lignite. In [30] it is reported that addition of sulphur to a nickel–molybdenum catalyst did not conclusively improve conversion of the Indonesian lignites being studied. In [31] Banko coal from Indonesia is hydrogenated in tetralin with a pyrite catalyst in the presence and absence of an amine. As an example of the findings, *n*-hexylamine at 0.4 g per g lignite raised the conversion from 65% to > 80%. Several other amines were used, and it was found that the chain length of the amine had an influence on the pre-asphaltene and asphaltene proportions of the liquid product.

The work in the following row, from almost 35 years ago, was the precursor to a great deal more on the hydrogenation of Victorian brown coals. It has been chosen for inclusion here because of its fundamental nature: hydrogenation tests of a set of coals of different lithotype (see section 2.4). Five different brown coal samples from Morwell were hydrogenated in tetralin at 380°C with a cobalt molybdate catalyst. Conversions were consistently 80%, with no lithotype dependence. This was at that time seen as a positive finding, as it opened the way to hydrogenation of coal as mined without regard to lithotype. Loy Yang coal, also from Victoria's Latrobe Valley, features in the next row. The work in [33] is strictly outside the scope of the table as there was no solvent, only hydrogen gas at 6 MPa, 400°C. Two catalysts were used, one nickel–molybdenum and the other sodium–nickel–molybdenum. Reacting times were either 30 minutes or 60 minutes. Conversions in the range 72–96% are reported [33]. Asphaltene yields were up to 25% of the original coal weight.

It is clear from the limited work reviewed above that catalyst and petrographic composition are influential in donor solvent hydrogenation of lignites. There is a great deal of related work in the recent literature. Realisation of the R&D by way of liquid fuel manufacture by this means is hoped for and will depend on patterns in oil prices.

14.5.3 Hydrogenation with carbon monoxide and water

A CO/H_2O mixture is an alternative to hydrogen with or without a 'shuttle' in the hydrogenation of lignites. The process is not new: it is stated in [35] that it was developed by Fischer in 1921. Returning to reference [30], additionally to treatment with hydrogen and a solvent, hydrogenation with CO/H_2O was carried out. Conditions were 3 MPa of carbon monoxide and temperatures of 365 or 400°C applied to a 35 cm^3 reactor the initial contents of which were coal, $NaAlO_2$ catalyst and liquid water. Final reactor conditions were sustained for 30 minutes. Across the range of coals in the study conversions of 60–70% were attained in this way.

A mechanistic account of the process is given in [26]. It consists partly of the water gas shift reaction:

$$CO + H_2O \rightarrow CO_2 + H_2$$

producing hydrogen for reaction with the lignite. There is also reduction of carbonyl groups in the coal by the carbon monoxide and generation of atomic hydrogen by reaction between the carbon monoxide and water in the coal structure.

14.6 Other activity in liquid fuel production

Table 14.3 gives details of some further investigations into the production of liquid fuels from lignites. They differ from examples previously covered in the use of methane and of synthesis gas respectively. Each is on a laboratory scale.

Table 14.3 Investigations into liquid fuel production from lignites.

REFERENCE	DETAILS
[36]	Leigh Creek coal (see section 10.3) treated with methane at 5 MPa, temperature up to 400°C, without a catalyst, with an aluminophosphate catalyst and with such a catalyst also incorporating lead. Aluminophosphate alone the more effective catalyst in terms of liquid yield
[37]	A Victorian brown coal (Morwell) treated with tetralin in a synthesis gas (CO/H_2) atmosphere at 405°C, reacting pressure up to 24 MPa. Several catalysts examined. Conversions up to 94% of the coal weight

In the work on Leigh Creek coal in the first row the most marked effect of the catalysts was on the hydrogen distribution in the liquid, either catalyst causing a reduction in the proportion of aromatic hydrogen. In the work on Morwell coal in the next row there was some emphasis on catalysis. Copper, nickel or cobalt were added to the coal as ions in solution. Aluminium was added as sodium aluminate and silicon as sodium silicate. Magnesium acetate was also added. Copper, nickel or cobalt with sodium aluminate was the most powerful at increasing yield.

14.7 Concluding remarks

The methods with which the chapter opened – F-T and Bergius – have not become obsolete, though they have been followed by procedures involving solvents. Neither the lignite nor the expertise to provide huge amounts of liquid fuels from it is lacking. All of the major oil companies have a programme into coal-to-liquid fuels. Additionally to the EDS coal liquefaction process there is activity by Shell, a senior official of which has said, in relation to Shell's endeavours in this area, 'that while the company had proven the technology works, the economic viability of such projects is not guaranteed' [38].

14.8 References

[1] *Fischer-Tropsch Fuels*, National Energy Technology Laboratory, Albany, OR (2011)

[2] http://www.fischer-tropsch.org/primary_documents/gvt_reports/BIOS/bios_1697.htm

[3] http://www.coaltooil.com/. See also http://www.usmexenergy.com/

[4] http://www.fischer-tropsch.org/primary_documents/gvt_reports/BIOS/bios_1697.htm

[5] Davis B.H., Occelli M.L. (eds) *Fischer-Tropsch Synthesis, Catalysts and Catalysis*, CRC Press (2009)

[6] http://www.scribd.com/doc/20294935/Fischer-Tropsch-Synthesis#scribd

[7] de Klerk A. *Fischer-Tropsch Refining*, John Wiley (2012)

[8] Braun A. *The German Economy in the Twentieth Century*, 1st edition (1990)

[9] http://tera-3.ul.cs.cmu.edu/NASD/4dcb85c3-9fee-4c83-9e6d-fe6ce5522b59/China/disk4/75/75-3/31005882/HTML/00000062.htm

[10] http://www.chemistryviews.org/details/ezine/5466751/100th_Anniversary_Bergius_Process.html

[11] http://www.nobelprize.org/nobel_prizes/chemistry/laureates/1931/bergius-bio.html

[12] http://memim.com/brabag.html

[13] *The Engineer* 1 April 1938, p. 365

[14] http://www.ausairpower.net/APA-USAF-SynFuels.html

[15] Shulyakovskaya L.V. 'Quantitative characteristics of hydrogenation reactivity of macerals of lignite from Kansk-Achinsk field – coals of the Itatskoe deposit', *Fuel and Energy Abstracts* 38: 5 (1997)

[16] Kusy J., Andel L., Safarova M., Vales J., Ciahotny K. 'Hydrogenation process of the tar obtained from the pyrolysis of brown coal', *Fuel* 101: 38–44 (2012)

[17] http://member.zeusintel.com/ZSRR/news_details.aspx?newsid=33009

[18] https://coalactionnetworkaotearoa.wordpress.com/lignite/

[19] Neavel R.C. 'Exxon donor solvent liquefaction process', *Philosophical Transactions of the Royal Society of London* A300: 141–156 (1981)

[20] Schlosberg R.H. *Chemistry of Coal Conversion*, Springer (1985)

[21] Shui H., Cai Z., Xu C. 'Recent advances in direct coal liquefaction', *Energies* 3: 155–170 (2010)

[22] Karacaa H., Ceylana K., Olcayb A. 'Catalytic dissolution of two Turkish lignites in tetralin under nitrogen atmosphere: effects of the extraction parameters on the conversion', *Fuel* 80: 559–564 (2001)

[23] Sharypov V.I., Kuznetsov B.N., Beregovtsova N.G., Reshetnikov O.L., Baryshnikov S.V. 'Modification of iron ore catalysts for lignite hydrogenation and hydrocracking of coal derived liquids', *Fuel* 75: 39–42 (1996)

[24] Wang W., Shui H., Zhang D., Gao J. 'A comparison of FeS, FeS + S and solid superacid

catalytic properties for coal hydroliquefaction', *Fuel* 86: 835–842 (2007)

[25] Cebolla V.L., Martínez M.T., Miranda J.L., Fernández I. 'Effects of petrographic composition and sulphur in liquefaction of Spanish lignites', *Fuel* 71: 81–85 (1992)

[26] Schobert H.H. *Lignites of North America*, Elsevier (1995)

[27] http://www.wvcoal.com/research-development/penn-state-writes-coaltl-qbookq.html

[28] http://www.britannica.com/EBchecked/topic/354359/maceral

[29] Ahmad A., Chaudry M.N., Shabbir A., Ali M. 'Thar coal, Sindh, Pakistan: a potential candidate for liquefaction', *Pakistan Journal of Hydrocarbon Research* 23: 41–49 (2013)

[30] Artantoa Y., Jackson W.R., Redlich P.J., Marshall M. 'Liquefaction studies of some Indonesian low rank coals', *Fuel* 79: 1333–1340 (2000)

[31] Arso A., Lino M. 'Effect of the addition of *n*-alkylamines on liquefaction of Banko coal', *Fuel Processing Technology* 85: 325–335 (2004)

[32] Hatswell M.R., Hertan P.A., Jackson W.R., Larkins F.P., Marshall M., Rash D. 'Direct hydrogenation of Victorian brown coal lithotypes', *Fuel* 60: 544–545 (1981)

[33] Hulston C.K.J., Redlich P.J., Jackson W.R., Larkins F.P., Marshall M. 'Nickel molybdate-catalysed hydrogenation of brown coal without solvent or added sulphur', *Fuel* 75: 1387–1392 (1996)

[34] *Fuel and Energy Abstracts* January 1995, p. 14

[35] Cassidy P.J., Jackson W.R., Larkins F.P., Sakurovs R.J., Sutton J.F. 'Hydrogenation of brown coal. 8. The effect of added promoters and water on the liquefaction of Victorian brown coal using hydrogen, carbon monoxide and synthesis gas', *Fuel* 65: 374–379 (1986)

[36] Yang K., Batts B.D., Wilson M.A., Gorbaty M.L., Maa P.S., Long M.A., He S.X.J., Attalla M.I. 'Reaction of methane with coal', *Fuel* 76: 1105–1115 (1997)

[37] Cassidy P.J., Jackson W.R., Larkins F.P., Sakurovs R. 'Hydrogenation of brown coal. 10. Multi-element promoters for the liquefaction of wet Victorian brown coal in synthesis gas', *Fuel* 65: 1057–1061 (1986)

[38] Linda Cook, quoted in: http://www.sourcewatch.org/index.php?title=Coal-to-Liquids

CHAPTER 15
CHEMICAL SUBSTANCES FROM LIGNITES

15.1 Introduction

The gasification processes covered in Chapter 13 also produce liquids, which have been put to use. The use of supercritical fluids to obtain extracts from lignites is an important further means of obtaining chemicals. In this chapter discussion of that will be preceded by a section on Montan wax.

15.2 Montan wax

15.2.1 Production

This is extracted from lignite, and a case study involving an Irish lignite will be given [1]. Montan wax was extracted from a Co. Antrim lignite using toluene, xylene and methanol. The best results – wax in a quantity of > 10% the coal weight – were obtained with toluene at 90°C. With a charge of 1 kg of the lignite, the extraction time was three hours.

Montan wax is manufactured in countries including Germany, where it is made from lignite originating at Amsdorf [2].

Germany is the major producer of Montan wax, which it exports to countries including Australia. To the example from Ireland above will be added accounts of a number of other research endeavours into Montan wax production. A number of Greek lignites including one from Megalopolis (see section 5.3.4) were examined for Montan wax yield [3] using various solvents and solvent mixtures. Those from the Megalopolis coal treated with benzene had a melting point in the range 64–70°C, whilst those from the same coal extracted in benzene/methanol had a melting point in the range 77–83°C. Comparisons are made with Montan waxes from other lignites. For example, a Montan wax from an Arkansas lignite had a melting point in the range 80–83°C and one from a Czech lignite had a melting point of 73–75°C with benzene-only extraction in each case. The corresponding values for extraction in benzene/methanol are 96–103°C and 72–75°C.

Two Turkish lignites were tested for Montan wax production by extraction work described in [4]. One of them, from Elbistan, gave a yield of about 1.5% of wax when Soxhlet-extracted in benzene. On the basis of solubility criteria the products were resolved into resin, asphalt and pure wax, respectively 20.1, 2.0 and 77.9% of the weight of extracted substance. Solvents other than benzene were used, for example toluene, which gave a product distribution of resin 11.7%, asphalt 19.3% and pure wax 69.0%, so when the solvent was changed for a similar one the effects on product distribution were quite major. Refining of Montan wax involves removal of the resin and the asphalt [5]. In evaluating a particular Montan wax for use, whether it blends satisfactorily with paraffin wax made from crude oil is a factor (see following section).

15.2.2 Uses

A great deal of Montan wax finds its way into polishes, although there are some applications much less general than that requiring precise product specification. These include lubricants for machinery. A novel application is to wood preservation. Boron compounds are used as wood preserving agents, and there have been difficulties due to loss of these from the wood by leaching under wet conditions. It is reported in [6] that this difficulty can be mitigated if the boric acid preservative as applied is not simply in aqueous solution, the procedure commonly followed, but in a Montan wax emulsion. Montan waxes have also been evaluated for protecting wood from termites [7]. Esters made from Montan wax by treatment with an alcohol have been used as a food additive, being applied to the surface of fresh fruits [8].

The next application to be described involved beads made from 25% Montan wax and 75% paraffin wax. Beads of this composition were applied to manufacture of reinforced concrete for road and pavement construction [9]. A difficulty had been encountered whereby salt used in de-icing had permeated the concrete and contacted steel reinforcement bars, hastening their corrosion. Additional to this effect *per se* is the fact that the steel on corrosion expands, leading to a spalling effect of the concrete. In the work described in [8] the composite Montan–paraffin wax in bead form was incorporated into the concrete. After the concrete had been cured it was heated to a temperature slightly above the melting point of the wax beads, the liquid so resulting closing the route which would have been taken by the salt solution to the steel bars which are thereby protected from corrosion. A similar application [10] is the possible use of Montan wax as the base of a grout material for containment barriers at hazardous waste sites.

It is clear then that there have been and continue to be important uses to which Montan wax can be put. Even in these days of word processors, 'carbon paper' has not completely disappeared (35 years ago or less, no office would have been without it). Montan wax is sometimes used in the manufacture of that, in one instance in the USA with Montan wax imported from Germany [11].

15.3 Use of supercritical fluids

15.3.1 Introduction

Supercritical fluids were explained in an appendix to Chapter 4 with reference to steam. A major branch of chemical technology is use of supercritical fluids in the extraction of chemicals. This as it relates to lignites is summarised below with suitable examples.

15.3.2 Examples

Table 15.1 gives some examples, and comments follow below. Here again an endeavour has been made to include lignites from sources having been referred to previously in the book.

Table 15.1 Supercritical extractions involving lignite.

REFERENCE	DETAILS
[12]	Four lignite samples from Soma, Turkey (see section 5.6) with and without prior demineralisation. Water at 400°C, 235 bar; at 450°C, 330 bar; at 500°C, 405 bar. 2 g samples in a 106 cm^3 vessel. Conversions of ≈ 35–40% at 400°C, declining to 30–35% at 500°C. Gas yields up to 660 cm^3
[13]	Supercritical toluene extraction from four Turkish lignites. Liquid yields of ≈ 10–20%. Yields raised by pre-treatment such as swelling in pyridine or removal of metal ions by acid exchange
[14]	Supercritical carbon dioxide extraction from Beulah (ND) lignite
[16]	Lignite from Dayan, Inner Mongolia. Extraction with supercritical water at 24 MPa and 30 MPa, temperatures up to 550°C. Conversions (gas + liquid) of up to 54%. Extract distributed between oil, asphaltenes and pre-asphaltenes
[18]	Five Victorian brown coals subjected to supercritical water. See discussion in the main text
[19]	Spanish lignite. Extraction from 50 g samples with supercritical methanol and supercritical ethanol. Critical points: Methanol 239°C, 8.1 MPa. Ethanol 241°C, 6.3 MPa. Temperature range with methanol in [19] 250–350°C. Pressure range 8.6–15.2 MPa. Same temperature range with ethanol, pressure range 7.1–15.2 MPa
[20]	Hungarian lignites treated in supercritical toluene at a range of pressures and temperatures. At 400°C, 16 MPa extract yields 25–30%
[21]	Two Turkish lignites – Gayniik and Tuncbilek – treated in supercritical water at 550°C, 30 MPa. Conversion 68% for both
[22]	Liptobiolith coal of lignite rank originating in Russia. Treatment with supercritical water in the temperature range 300–470°C. Up to 23% extract yield

One can easily confirm from the figure in the appendix to Chapter 4 that the temperature–pressure pairs in the first row correspond to the supercritical state. One factor in the decline of conversion with increasing temperature is believed to be polymerisation of the carbon structure under the more vigorous extraction conditions. In other words, the decline in conversion is better understood as an increase in the carbon content of the residue. Gaseous product, the yield of which is given in the table, comprised $C_{1\rightarrow 4+}$ hydrocarbons as well as carbon monoxide and hydrogen. The second row is also concerned with Turkish lignites, with toluene as the fluid in the supercritical state. The critical temperature of toluene is 319°C, and in [13] the toluene was just into the supercritical region at 330°C. In the work summarised in the following row supercritical carbon dioxide was the extracting fluid, at a pressure of 40.5 MPa. A reader can consult the phase diagram for carbon dioxide [15] to deduce how far above the critical pressure of carbon dioxide this is. Applications of the lignite from Dayan (next row of the table, and also referred to in section 13.9.3) include underground gasification [17].

Victorian brown coals feature in the next row. Three of them were from open cut mines having already featured in this text, Yallourn, Morwell and Loy Yang. One was from Gelliondale and one from Coolungoolun; these coals are not used in electricity generation. The Gelliondale coal was tested at 22 MPa – close to the critical pressure – at temperatures in the range 380–460°C. Conversion was ≈ 55% and extract yields ≈ 22%. Oils exceeded asphaltenes and pre-asphaltenes in the liquid product. Under conditions very close to the critical point for water, the Loy Yang coal gave 51% conversion, the Coolungoolun coal 42%, the Yallourn coal 54% and the Morwell coal 82%. The remarkably high conversion for the Morwell coal was roughly half-and-half gas and extract. Moving to the next row where the supercritical fluid is methanol or ethanol, we note that the wider pressure range was possible with ethanol because of its lower critical pressure compared with methanol. The point is made in [19] that when the extracting liquid is organic it can add to the carbon content of the product, and anomalous mass balances exceeding 100% are observed. By contrast, when Victorian brown coals are treated with supercritical water (previous row) mass balances tend to close below 100% because of unaccounted for moisture losses from the coal.

The work in reference [21], in common with that in reference [16], uses sub-critical water as well as supercritical and a brief comparison is helpful. We recall that the critical point for water is 374°C, 21.8 MPa making the conditions given in the table supercritical. At the same temperature and at a pressure of 2.5 MPa the conversion is 55.5%. The term liptobiolith coal (final row) was introduced in section 5.1.2. Products from the supercritical extraction of this Russian example included asphaltenes and resins [22]. These had molar masses respectively up to ≈ 1000 g and ≈ 300 g.

15.4 Chemicals from solvent extraction

The title of this section means of course extraction by solvents under sub-critical conditions. Table 15.2 gives details of studies of solvent extraction of chemicals from lignites. The volume of literature on this is huge, and more recent work has been selected. Extraction from a Brandon lignite, discussed in section 6.11, could equally appropriately have been included in this section.

Table 15.2 Solvent extraction from lignites.

LOCATION AND REFERENCE	DETAILS
Hulunbeier, Inner Mongolia [23]	Extraction from 2 g samples with carbon disulphide, acetone and ethanol. GC-MS analysis. Compounds in the CS_2 extract largely aliphatic up to about C_{30}. The most abundant single compound $C_{28}H_{58}$ at 6.2%. Oxygenated compounds exceeding aliphatic hydrocarbons in amount in the acetone extract. Major aliphatics and oxygenates in the ethanol extract. Small amounts of heterocycles with all three solvents
Loy Yang, Victoria [24]	Extraction by solvents including 1-methyl naphthalene (1-MN). See comments in the main text
Xianfeng lignite [25]	Extraction into a toluene–methanol mixture. Carbon content of the extract 77.7% compared with 63.1% for the parent coal. The extract treated with tetrahydrofuran (THF). The THF-soluble part separated into fractions in a column of silica gel and eluted
Shengli lignite [26]	Extraction in 1-MN and mixtures of 1-MN with polar solvents at 360°C. Yield of 30% of the coal weight with 1-MN only, increasing to 81% with 1-MN–15% ethanol
Spanish lignite [27]	Extraction of humic substances into potassium hydroxide solution

In work on Loy Yang coal (row 2 of the table) fractions of coal were obtained by use of different extraction temperatures in 1-MN: 200–250°C, 250–300°C, 300–350°C and > 350°C. The carbon content of the coal before extraction was 66.9%, and this increased over the fractions to 76.8% for those at the highest temperatures. Irrespective of any value of the chemicals extracted, this represents beneficiation of the coal. The increase in carbon content is obviously accompanied by the presence of oxygenated substances in the extracts. At the highest temperatures the yield of extracted material was 45.3% of the coal weight.

Moving to the next row, the fractions obtained when the THF-soluble part of the extract from the Xianfeng lignite was eluted had carbon contents in the range 50.8–82.7%. Compounds identified on analysis included aliphatics up to C_{30} and aromatics up to C_{27}. Amongst the oxygenated substances were esters. The model of coal as a polymer is applied to the work on the Shengli lignite in the following row. Between 360 and 380°C with 1-MN only as the solvent there is a dramatic increase in the yield of extract from 30% of the coal weight as noted to 81% and this is attributed to loss of the polymer structure. An alternative view would be that it is due to breakdown of the cross-linked part of the coal. Shengli lignite features later in the chapter. In the extraction of humic acids into an aqueous medium (final row), yields, expressed as a percentage of the total carbon content, were up to 80%.

It is clear from the very limited number of studies examined that points of interest and possible application in solvent extraction of lignites are the miscellany of organic compounds produced and the high carbon content of the solid residue.

15.5 Specialised products

15.5.1 Drilling fluid additives

Specialised products from lignites are many, and include fluid loss additives for drilling fluids used at oil fields [28]. A difficulty with the use of drilling fluids is their loss to the outside of the drill tubes, and the use of lignites in the fluid has been found to control this. This requires selection of lignite from close to an exposed surface at the mine, so that there has been some oxidation to acids [29]. Alternatively it is spread out so as to oxidise in the atmosphere. The action of the lignite as a fluid control additive depends on its ability to dissolve, and this is brought about by alkali which will enable the acid content referred to to enter the aqueous phase, and one criterion for the acceptability of a lignite additive to an aqueous drilling fluid is that there should be 75% dissolution [29]. Sometimes lignite for such use is mixed with caustic soda for this reason, being known as causticised lignite.

An example of causticised lignite is GLO CLZ 1000, a product of Global Drilling Fluids and Chemicals Ltd. in Gujurat, India [25]. It is of particle size such that its bulk density is 900–1000 kg m^{-3}. Amounts in which it is added to a drilling fluid depend on requirements, as it not only affects fluid loss as noted but is also a factor in the viscosity and the operating temperature range. In [30] it is recommended that it is added in amounts in the range 2.85–28.5 kg per cubic metre of drilling fluid. Prior dispersion in water alone is sometimes necessary.

Other lignite products are made available by Global Drilling Fluids and Chemicals Ltd. One is plain lignite powder, coded GLO LIG 1000, and it is clear from the opening paragraph of this section that this is closer to being a norm than a causticised product. Another is potassium lignite GLO KLIG 1000 which by introducing potassium ions into the drilling fluid influences its rheology [31].

Fluid loss control additives for drilling fluids can also be made from lignite which has been treated with an amine. Having regard to the fact that amines are bases, their action on lignites is broadly similar to that of alkalis. In view of what was said in section 6.10 about leonardite – that it has undergone oxidation in the bed – it is not surprising that this also has been used in drilling fluids. Imprecision of word usage is evident in related literature, and leonardite for such use is sometimes referred to as 'lignite'. Coal near the surface of a lignite mine displaying some oxidation has also been referred to as 'leonardite'. There is much more on this in Chapter 18.

15.5.2 Carbon electrodes

A notable example of production of carbon electrodes with a lignite starting material is Rockdale TX [32]. Production there began in preparation for the Korean war. The US

Air Force needed more planes. Aluminium production for their construction materials required carbon electrodes in large volume. There have been investigations into electrode production from Loy Yang coal in Victoria's Latrobe Valley [33]. It is noted in section 10.2 that there has been aluminium production in Victoria.

15.5.3 Jet

'Jet' or 'lignite jet' is a form of lignite [34], and frequently finds application as the basis of a gemstone. In England it occurs in a deposit in East Yorkshire, and commercially this is one of the world's major sources. There are also reserves in countries including Spain, the USA, Sweden and Germany.

The process of jet formation over geological time is called 'jetification'. In laboratory simulations on wood [35] it has been shown that humid and oxygen-rich conditions in which there is also exposure to light promote the transformation to a shiny, black material and this can be taken to represent events leading to the formation of jet deposits. Exposure to light is believed to be particularly important. Table 15.3 gives some properties of jet samples.

Table 15.3 Properties of jet.

PROPERTY	VALUES AND REFERENCES
Density	1.32 g cm^3 at typical value [35]
Hardness	2.5–4 on the Mohs scale [37]
Refractive index	1.64–1.68 [39]

The density in row 1 of the table is within the range of true densities, a.k.a. helium densities because they are determined by displacement of helium, of lignites. Schobert [36] gives values for Darco (PA), Zap (ND) and Harmon (ND) lignites of respectively 1.30, 1.40 and 1.45 g cm^{-3}. That oxygen-rich conditions, promoting jetification as noted, would be accompanied by compression is counterintuitive, so that jet has about the same density as lignite is not surprising. The hardness range of jet on the Mohs scale can be compared with 1.25 for gypsum, 7 for granite and 10 for diamond [38]. Amongst materials used in making jewellery, pearl has a Mohs hardness of 2.5–3.5, tortoiseshell 2.5 and ivory 2–3 [39,40]. Jet has a lower thermal conductivity than most such materials.

Siberian lignite featured centrally in the part of the book concerned with power generation in the FSU. Jet occurs in eastern Siberia, where it has also been classified as sapropelic coal [41] consistently with the absence of a banded structure. There is jet in Tibet, and significant trade in artefacts made from it [42].

15.6 Treatment with oxidising reagents

That there is current interest in chemicals from lignites via oxidation with hydrogen peroxide is evidenced by recent major publications on the topic, a selection of which will be reviewed here. Carboxylic acids were the dominant class of product when a Chinese lignite was reacted with hydrogen peroxide and acetic anhydride at 50°C [43]. Malonic acid ($CH_2(COOH)_2$) and succinic acid ($(CH_2COOH)_2$) were by far the dominant products, accounting for over 40% of the total carboxylic acid yield. It was concluded that these originated from $-CH_2-$ and $-CH_2CH_2-$ linkages in the lignite structure. Malonic and succinic acids were also dominant products when a lignite from Moravia was oxidised with hydrogen peroxide [44] at temperatures in the range 30–70°C. Selectivity of the process to these products is noted in [44]. The lowest temperature led to the highest yield of malonic acid, and the proportions of malonic to succinic were strongly dependent on reacting temperature.

A lignite from Shengli was oxidised with hydrogen peroxide in work described in [45], in which tests with sodium hypochlorite as oxidant were also carried out as was ruthenium ion-catalysed oxidation, in which the oxidant is sodium periodate, $NaIO_4$. Ruthenium ion-catalysed oxidation is seen as a route to production of organic chemicals from lignite on a larger scale. With ruthenium ion-catalysed oxidation, succinic acid hugely dominated the products. (On a large scale, the high cost of ruthenium would be a factor in the viability of its use as a catalyst.) With hydrogen peroxide there were in the oxidation products carboxylic acids containing methoxy and hydroxyl groups in the carbon structures. Oxidation with sodium hypochlorite yielded carboxylic acids with more than one carboxyl group on a benzene ring, for example benzene tetracarboxylic acids, $C_6H_2(COOH)_4$.

15.7 Use of lignite fly ash in carbon sequestration

It was mentioned in previous sections including 5.2.3 that fly ash from lignite combustion does sometimes find a use, for example in brick manufacture. A further use is in carbon sequestration, when the fuel from which the carbon dioxide comes may or may not have been a lignite but in general will not have been. That will be discussed in this chapter, where the fly ash will be viewed as a 'chemical substance' for subsequent use obtained from the lignite. Carbon capture and sequestration from lignite fuels are the subject of Chapter 19. Some such applications of fly ash from lignite are given in Table 15.4, and comments follow below. Obviously an important part is played by the calcium oxide and magnesium oxide contents of the fly ash, which can react with carbon dioxide to form carbonates, a process sometimes described as carbonation.

Table 15.4 Lignite fly ash in carbon sequestration.

REFERENCE	DETAILS
[46]	Fly ash from three Polish lignite-fired power plants: Turów, Bełchatów and Pątnów (see section 5.2.3 et seq.). CO_2 uptakes of respectively 8.81, 4.12 and 7.87 g per 100 g fly ash after 240 hours of exposure of the CO_2 to a shaken aqueous suspension of the fly ash
[47]	Carbonation of fly ash from the burning of a Rhenish lignite (see section 4.2.3). Fly ash made into a slurry with water. Uptakes of up to 4.8 mmol per g of fly ash (21 g per 100 g fly ash)
[48]	Fly ash from Victorian brown coal. Aqueous conditions. Conversion of CaO to Ca^{2+} with acetic acid. Formation of magnesium bicarbonate – $Mg(HCO_3)_2$ – as an intermediate in the formation of magnesium carbonate. Uptakes up to 264 kg CO_2 per tonne of fly ash (26.4 g CO_2 per 100 g fly ash)

In the work in the first row of the table, shaking of the fly ash suspension was found to be a major factor in the extent of carbon dioxide uptake. About 6 million tonnes of fly ash are captured annually in the lignite-fired power plants of Poland [42], so using a value of say 6 g of CO_2 per 100 g fly ash the potential for sequestration is 0.36 million tonnes, a calculation which a reader can easily confirm for him/herself. In the work described in the second row the slurry was stirred, and there was a dependence of the carbonation performance on the stirrer speed as well as on the solids loading of the slurry. The fly ash used in the work in the third row was taken from the Hazelwood power station (see section 10.2.2).

15.8 Concluding remarks

This chapter has covered chemicals production by extraction, and extended that to oxidation as well as to drilling fluids and to carbon electrodes. This chapter was also seen as the natural place to discuss lignite jet. What the chapter has not been concerned with is chemicals production via synthesis gas. This of course was the subject of Chapters 13 and 14, and the 'chemicals' so produced are liquid fuels or products other than fuels from synthesis gas. An example of manufacture of an important product other than liquid fuels obtained from synthesis gas itself made from lignite will conclude this chapter. In wartime Germany, the manufacture of nitrile rubber ('Buna') from lignite-derived synthesis gas by the German giant I.G. was next only in importance to the production of aviation fuel from the same source. More details can be found in [49].

15.9 References

[1] McKay G., Alien S.J., McGookin C. 'Montan wax extraction from Northern Ireland lignite', *Fuel* 67: 1703–1705 (1988)

[2] http://www.wachs-und-mehr.de/index.php/en/unternehmen

[3] Kampouris E.M., Constantinoglou C., Prounias N., Sideropoulos T. 'Extractable waxes from Greek lignites', *Fuel* 52: 47–51 (1973)

[4] Gurkan T., Gurer S. 'Investigation of parameters affecting atmospheric extraction yields and extract qualities of two Turkish lignites', *Fuel Processing Technology* 10: 19–31 (1985)

[5] Wei X., Yuan C., Zhang H., Li B. 'Montan wax: the state-of-the-art review', *Journal of Chemical and Pharmaceutical Research* 6: 1230–1236 (2014)

[6] Lesar B., Kralj P., Humar M. 'Montan wax improves performance of boron-based wood preservatives', *International Biodeterioration & Biodegradation* 63: 306–310 (2009)

[7] Scholz G., Militz H., Gascón-Garrido P., Ibiza-Palacios M.S., Oliver-Villanueva J.V., Peters B.C., Fitzgerald C.J. 'Improved termite resistance of wood by wax impregnation', *International Biodeterioration & Biodegradation* 64: 688–693 (2010)

[8] European Food Safety Authority (EFSA) Panel on Food Additives and Nutrient Sources added to Food, Scientific Opinion on the re-evaluation of Montan acid esters (E 912) as a food additive. *EFSA Journal* 11 (2013), 21pp.

[9] Geishauser C.B., Cady P.D. 'A study of the heat treating cycle for internally sealed concrete containing Montan–paraffin wax beads', *Cement and Concrete Research* 7: 85–94 (1977)

[10] *International Journal of Rock Mechanics and Mining Sciences & Geomechanics Abstracts* 33: 173 (1996)

[11] http://www.strohmeyer.com/montanwax.html

[12] Sert M., Ballice L., Yüksel M., Saglam M. 'Effect of mineral matter on product yield in supercritical water extraction of lignite at different temperatures', *Journal of Supercritical Fluids* 57: 213–218 (2011)

[13] Simsk E.H., Karaduman A., Caliskan S., Togrul T. 'The effect of pre-swelling and/or pre-treatment of some Turkish coals on the supercritical fluid extract yield', *Fuel* 81: 503–506 (2002)

[14] Louie P.K.K., Timpe R.C., Hawthorne S.B., Miller D.J. 'Determination of elemental sulphur in coal by supercritical fluid extraction and gas chromatography with atomic emission detection', *Fuel* 72: 225–231 (1993)

[15] http://commons.wikimedia.org/wiki/File:Carbon_dioxide_pressure-temperature_phase_diagram.svg

[16] Hu H., Guo S., Hedden K. 'Extraction of lignite with water in sub- and supercritical states', *Fuel Processing Technology* 53: 269–277 (1998)

[17] Zhanga H., Zhoua X. 'Speciation variation of trace metals in coal gasification and combustion', *Chemical Speciation and Bioavailability* 21: 93–97 (2009)

[18] Kershaw J.R. 'Extraction of Victorian brown coals with supercritical water', *Fuel Processing Technology* 13: 111–124 (1986)

[19] Garcia R., Moinelo S.R., Snape C.E. 'A spectroscopic study of structural changes occurring in supercritical gas extraction of a Spanish lignite with alcohols', *Fuel* 72: 427–433 (1993)

[20] *Fuel and Energy Abstracts* May 1998, p. 168

[21] Canel M., Missal P. 'Extraction of solid fuels with sub- and supercritical water', *Fuel* 73:

1776–1780 (1994)

[22] Fedyaeva O.N., Vostrikova A.A., Shishkina A.V., Fedorovab N.I. 'Liquefaction of lipto-biolith coal in supercritical water flow under non-isothermal conditions', *Russian Journal of Physical Chemistry B* 8: 1054–1063 (2014)

[23] Tahmasebi A., Jiang Y., Yu J., Li X., Lucas J. 'Solvent extraction of Chinese lignite and chemical structure changes of the residue during H_2O_2 oxidation', *Fuel Processing Technology* 129: 213–221 (2015)

[24] Ashida R., Morimoto M., Makino Y., Umemoto S., Nakagawa H., Miura K., Saito K., Kato K. Fractionation of brown coal by sequential high temperature solvent extraction, *Fuel* 88: 1485–1490 (2009)

[25] Wang Z., Shui H., Pan C., Li L., Ren S., Lei Z., Kang S., Wei C., Hu J. 'Structural characterization of the thermal extracts of lignite', *Fuel Processing Technology* 120: 8–15 (2014)

[26] Shui H., Zhu W., Wang W., Pan C., Wang Z., Lei Z., Ren S., Kang S. 'Thermal dissolution of lignite and liquefaction behaviors of its thermal dissolution soluble fractions', *Fuel* 139: 516–522 (2015)

[27] Garcia D., Cegarra J., Roig A. 'Effects of the extraction temperature on the characteristics of a humic fertiliser obtained from lignite', *Bioresource Technology* 47: 103–106 (1994)

[28] http://www.glossary.oilfield.slb.com/en/Terms.aspx?LookIn=term%20name&filter=lignite

[29] *Ven News* March 1991, Venture Chemicals, Lafayette, LA

[30] http://www.globaldrillchem.com/products/index/causticized-lignite_3853.html

[31] http://www.universaloilfield.org/Universal%20Oil%20FieldsB/RESINATED-LIGNITE.php

[32] http://www.usgwarchives.net/tx/milam/history/pg016.htm

[33] Gardner H.J., Waters P.L., Watts A. 'Production of electrode carbon from brown-coal char and gaseous hydrocarbons', *Preprints of Papers, National Conference on Chemical Engineering* pp. 80–85, Institution of Engineers Australia (1976)

[34] Dill H.G. 'The "chessboard" classification scheme of mineral deposits: mineralogy and geology from aluminium to zirconium', *Earth Science Reviews* 100: 1–420 (2010)

[35] Kool J., Poole I., van Bergen P.F. 'How jet is formed: an organic geochemical approach using pyrolysis gas chromatography–mass spectrometry', *Organic Geochemistry* 40: 700–705 (2009)

[36] Schobert H.H. *Lignites of North America*, Elsevier (1995)

[37] https://cycleback.wordpress.com/2013/02/11/a-tool-for-identifying-materials-the-mohs-scale-of-hardness/

[38] http://www.rauschgranite.com/resources/hardness-of-natural-stone/

[39] Read P.G. *Gemmology*, Elsevier (2013)

[40] http://www.gemdat.org/gem-9355.html

[41] 'Siberian sapropelic coal: a unique type of workable jet', *Fuel and Energy Abstracts* Sep-

tember 1995, p. 324

[42] http://www.crystalsrocksandgems.com/Healing_Crystals/Jet.html

[43] Liu J., Wei X-W., Wang Y-G., Zhang D-D., Wang T-M., Lv J-H., Gui J., Qu M., Zong Z-M. 'Mild oxidation of Xiaolongtan lignite in aqueous hydrogen peroxide–acetic anhydride', *Fuel* 142: 268–273 (2015)

[44] Doskoč L., Grasset L., Válková D., Pekar M. 'Hydrogen peroxide oxidation of humic acids and lignite', *Fuel* 134: 406–413 (2014)

[45] Liu F-J., Wei X-Y., Zhu Y., Gui J., Wang Y-G., Fan X., Zhao Y-P., Zong Z-M., Zhao W. 'Investigation on structural features of Shengli lignite through oxidation under mild conditions', *Fuel* 109: 316–324 (2013)

[46] Uliasz-Bocheńczyka A., Mokrzyckia E., Piotrowskib Z., Pomykałab R. 'Estimation of CO_2 sequestration potential via mineral carbonation in fly ash from lignite combustion in Poland', *Energy Procedia* 1: 4783–4789 (2009)

[47] Bauer M., Gassen N., Stanjek H., Peiffer S. 'Carbonation of lignite fly ash at ambient T and P in a semi-dry reaction system for CO_2 sequestration', *Applied Geochemistry* 26: 1502–1512 (2011)

[48] Sun Y., Parikh V., Zhang L. 'Sequestration of carbon dioxide by indirect mineralization using Victorian brown coal fly ash', *Journal of Hazardous Materials* 209: 458–466 (2012)

[49] Borkin J. *The Crime and Punishment of IG Farben*, Free Press (1978)

CHAPTER 16
UNWORKED LIGNITE DEPOSITS

16.1 Introduction

Some countries have lignite/brown coal reserves without ever having mined it to a significant degree. Details will be given, and comments on the prospects of development of such reserves in the future will be offered. 'Unworked' has to be understood as 'having never been the scene of major production'. So often at a deposit seen as unworked there has at some time been local use. This is emphasised below with Ecuador as an example. Greenland is used as an example of a country having in the distant past had moderate lignite production.

16.2 Examples

Table 16.1 gives some examples.

Table 16.1 Lignite deposits awaiting development.

LOCATION	DETAILS
Argentina	500 million tonnes of known reserves [1]. No current production [2,3]
Sweden	See comments in the main text
Chile	155 million tonnes of known reserves [1]. No current production [5]
Colombia	380 million tonnes of known reserves [1]
Ecuador	22 million tonnes of known reserves [1]
Japan	10–15 million tonnes of known reserves [1]
Egypt	22 million tonnes of known reserves [1]
Greenland	183 million tonnes of known reserves [1]
Nigeria	169 million tonnes of known reserves [1]. No current production
Israel	0.3 billion tonnes at Hula: estimated by the author from information in section 7.9

The Neuquen basin in Argentina is a major source of natural gas and also contains some lignite. This is also true of the Thrace basin (see section 13.9.3). The lignite reserves of Sweden are of the order of thousands of tonnes, not meriting investment, but the country has been included for the following reason. Vattenfall (see section 4.2.5 *et seq.*), which is very active in lignite-fired power stations in Germany and in Greece, is 100% owned by the government of Sweden. The government is now requiring the company not to expand its activity with lignite in Germany [4].

In Chile (following row) the Pecket deposit contains both lignite and sub-bituminous coal. This is quantified in [6] as vitrinite reflectance in the range 0.28–0.42%. Lignites in Colombia include those from Aránzazu. A relatively recent study of a lignite from this source [7] revealed a surprisingly high liptinite content, 26.2%, and a correspondingly low vitrinite content of 50.6%. The mineral content was predominantly pyrite and kaolinite ($Al_2Si_2O_5(OH)_4$).

Although the reserves of lignite in Ecuador (next row) are an order of magnitude lower than those of Argentina and Colombia they exceed those of Ethiopia where, as described earlier in the book, there is activity. An obvious point of contrast, influencing the viability of lignite usage, is that Ecuador, a member of OPEC, produces a great deal more oil than Ethiopia does; they rank respectively 29th and 114th in oil production on a world basis [8]. The coal reserves of Ecuador are sub-bituminous and lignite. There was over 50 years ago very minor production at the Azogue-Biblian lignite mine in Ecuador with hand, not mechanical, methods. Similarly lignite from the Loja basin in Ecuador has supplied local industry [9].

Japan (following row) has coal reserves across the ranks, but no longer mines coal, finding it better to import it from Australia and Indonesia. In the 21st century the most important point in relation to the lignite resources of Japan is the effect of earthquakes on abandoned lignite mines. The degree of resistance to earthquake damage of lignite mines in Japan has been correlated with the original method of mine construction [10]. Moving on to Egypt (next row), lignite occurs in Sinai and comprises $\approx$ 10 cm layers within kaolin, which is much more abundant than the lignite [11]. In Greenland lignite was mined at Qullissat up to about 45 years ago [12]. Note the high reserves, in spite of which Greenland imports in small quantities lignite briquettes. A picture of Qullissat forms Figure 16.1. Past lignite production in Iceland is discussed in section 23.2.

Figure 16.1 Qullissat, Greenland, once the scene of lignite mining. Courtesy of greenlandphotos.net

With reference to Nigeria (following row), lignites from Ogwashi–Asaba in the Anambra basin have recently been characterised [13]. Seven samples were found to vary in moisture content from 22.2% to 43.6% and in ash from 5.2% to an extraordinary 69.4%. Sulphur varied from too low to detect to 7.7%. Points of interest emerge from the petrographic analysis, including variation in macerals of the huminite group from 84% down to 10%. The sample having 10% huminite macerals had 87% of liptinite macerals. Huminite reflectances were in the range 0.24–0.36.The sample having very high ash had a mineral matter content in which clay was prevalent.

16.3 Evaluation of the utilisation potential of the unworked lignite

The figures in the right-hand side of Table 16.1 add up to a quantity of the order of 2000 million tonnes. This probably represents bed-moist lignite, so will be divided by two to give 1000 million tonnes of lignite suitable for burning, gasifying or converting to liquid fuels. It was stated in section 14.1 that in a F-T conversion one ton (equate to one tonne for a rough calculation) of lignite produces 3–4 barrels of liquid product. The 1000 million tonnes of lignite in the table could, if hypothetically converted to liquid fuel, realise about 4000 million barrels of liquid product. This is a long way short of being a staggering figure, equating to world needs for seven weeks. The Ghawar oil field in Saudi Arabia contains a quantity of crude oil an order of magnitude higher than this. Similarly, 4000 million tonnes of lignite could produce about 4×10^{19} J of electricity or about 11 200 TW-hour.

These figures have a negative suggestion about them, but in any case the word 'hypothetically' in the previous paragraph cannot be emphasised too much. Utilisation of all of the lignite to make liquid fuels or to generate electricity is impossible to conceive in realistic terms.

16.4 Concluding remarks

Lignites are seen as being heavily polluting fuel (e.g. [14–16]), although counter views have been expressed [17]. Obviously, when a fuel is seen as being 'dirty' (a word which recurs in the references cited) the method of burning it is at least part of the origin of the difficulty. Standards of combustion hygiene can be high or low with any fuel. A calculation to show that lignite produces more carbon dioxide per unit energy released than oil products and natural gas do is too elementary to merit reproduction here. Arguments against lignite usage on this basis disregard two factors: the advanced methods there are for CCS and, perhaps more importantly, the fact that any country has to balance high releasers of carbon dioxide with lower releasers and with non-releasers in its energy mix.

16.5 References

[1] http://data.un.org/Data.aspx?d=EDATA&f=cmID%3ALB%3BtrID%3A1511

[2] http://knoema.com/atlas/Argentina/topics/Energy/Coal/Production-of-Lignite-Coal

[3] http://www.helgilibrary.com/indicators/index/lignite-coal-production-as-of-worldwide-production/argentina

[4] http://treealerts.org/region/europe/2014/10/lausitz-lignite-expansion-in-doubt-as-swedish-government-signals-shift-to-renewables/

[5] http://treealerts.org/region/europe/2014/10/lausitz-lignite-expansion-in-doubt-as-swedish-government-signals-shift-to-renewables/

[6] Hidalgo E., Helle S., Alfaro G., Kelm U. 'Geology and characterisation of the Pecket coal deposit, Magellan Region, Chile', *International Journal of Coal Geology* 48: 233– 243 (2002)

[7] López I.C., Ward C.R. 'Composition and mode of occurrence of mineral matter in some Colombian coals', *International Journal of Coal Geology* 73: 3–18 (2008)

[8] https://www.cia.gov/library/publications/the-world-factbook/rankorder/2241rank.html

[9] http://digicoll.library.wisc.edu/cgi-bin/EcoNatRes/EcoNatRes-idx?type=turn&id=EcoNatRes.MinYB1963v4&entity=EcoNatRes.MinYB1963v4.p0307&isize=text

[10] Aydani O., Tano H. 'The damage to abandoned mines and quarries by the Great East Japan Earthquake on March 11th 2011', *Proceedings of the International Symposium on Engineering Lessons Learned from the 2011 Great East Japan Earthquake*, 1–4 March 2012, Tokyo, Japan, pp. 981–992

[11] Baioumy H. 'Hydrogen and oxygen isotopic compositions of sedimentary kaolin deposits, Egypt: paleoclimatic implications', *Applied Geochemistry* 29:_182–188 (2013)

[12] http://www.britannica.com/EBchecked/topic/485710/Qeqertarsuaq

[13] Ogala J., Siavalas G., Christanis K. 'Coal petrography, mineralogy and geochemistry of lignite samples from the Ogwashi–Asaba Formation, Nigeria', *Journal of African Earth Sciences* 66: 35–45 (2012)

[14] http://e360.yale.edu/feature/on_the_road_to_green_energy_germany_detours_on_dirty_coal/2769/

[15] http://www.bloomberg.com/bw/articles/2014-02-27/in-europe-dirty-coal-makes-a-comeback

[16] https://sunshinehours.wordpress.com/2014/08/28/europe-to-burn-dirty-filthy-lignite-instead-of-clean-fracked-natural-gas-or-nuclear/

[17] http://www.energypost.eu/wishing-away-lignite-eu-climate-policy-ignores-elephant-room/

CHAPTER 17
HAZARDS WITH LIGNITES

17.1 Introduction

The primary hazards are coal seam fires, dust explosions and spontaneous heating in storage. These will be dealt with in turn.

17.2 Seam fires at lignite mines

17.2.1 Examples

One of the best known examples of such a fire at a lignite deposit is that at Hoher Meissner in Hesse, Germany [1]. It is not very distant from other German lignite reserves discussed in this book; Frankfurt am Main, the principal city of Hesse, is 120 miles from Cologne. The fire at Hoher Meissner began centuries ago. A major difficulty with all coal seam fires is of course the combustion products, including sulphur dioxide, and their release into the atmosphere.

Another example of combustion at a lignite deposit is Smoking Hills, in Canada's North West Territory [2]. The fire was discovered about 90 years ago but is believed to have been burning in the early 19th century.

Smoking Hills is close to Franklin Bay, which receives water from the Franklin River. The water has become acidified through the sulphur dioxide emissions. In East Kalimantan, Indonesia there is a lignite seam fire and it is believed to have initiated a forest fire [3]. One might expect that a coal seam fire close to a forest would have a further harmful effect: few things are more damaging to tree life than acid rain which, of course, is formed from sulphur dioxide. This example of a lignite seam fire and the previous one both highlight the destructive potential towards the environment.

17.2.2 Further example: Hazelwood, Victoria, Australia 2014

The seam fires discussed in the previous section have been burning for very long periods and, with the exception of the one in East Kalimantan, have become part of the natural scenery and even tourist attractions. The fire at Hazelwood in 2014 was altogether different. That Hazelwood is the scene of lignite-fired electricity generation was stated in section 10.2.2. A fire at the mine began on 9 February 2014. After firefighting and extinguishment procedures it was handed back to the operators on 25th March [4].

It is widely held that ignition at Hazelwood was by an external heat source in the form of embers from a nearby fire [5]. The chief impact of the Hazelwood fire was of course on the atmosphere, and it is noted in [5] that ash particles were of particular concern in assessing community health effects.

17.2.3 Further comments

Laboratory simulation of a lignite seam fire is reported in [6]. It uses lignite from Ulanqab in Inner Mongolia. The lignite was crushed and dried, and 80 g samples placed in an electrically heated furnace at temperatures in the range 600–1000°C. Of particular interest was the formation of polyaromatic hydrocarbons (PAH), and these were found to be most abundant at a furnace temperature of 800°C. There was also a dependence of the PAH yield on the lignite particle size, larger particles yielding more PAH.

17.3 Dust explosions

17.3.1 Introduction

These can occur where lignite is being handled, and briquette factories are an obvious example. Smoking materials are in today's world not allowed in workplaces anyway, but when they were they were prohibited at briquette factories. In about the 1930s management at the briquette factory in Victoria, Australia (see section 10.2.2) declared that any employee caught smoking in the factory would not be merely dismissed but also prosecuted. (There is a return to Victoria in section 17.3.3.) The flammability limits of lignite in air are usually in the range 0.09–7 kg m^{-3} with a particle size dependence [7].

17.3.2 Literature survey

Table 17.1 gives details of some literature on dust explosions with lignites. Some background information is required. A dust explosion in air from any substance (susceptible materials include ground pharmaceuticals, sugar, wood and cork) is expressed in terms of two quantities. One is the maximum pressure P_{max}. The other is the index K_{st} defined by:

$$K_{st} = (dP/dt)_{max} \times V^{1/3}$$

where $(dP/dt)_{max}$ is the maximum rate of pressure rise and V the vessel volume. The units of K_{st} are bar m s^{-1} and a particular dust is after testing placed in one of the following categories [8]: ST class 0, K_{st} value = 0; ST class 1, K_{st} value less than 200 bar m s^{-1}; ST class 2, K_{st} value between 200 and 300 bar m s^{-1}; ST class 3, K_{st} value greater than 300 bar m s^{-1}. One international standard amongst a number for determination of P_{max} and K_{st} is BS EN 14034.

The K_{st} values in the first row of the table signify four samples at ST class 1 and one at class 2. Reference [8] also gives values of K_{st} for higher rank coals and for biomass. These include a Pittsburgh bituminous coal, for which K_{st} is 41 bar m s^{-1} and P_{max} 6.7 bar. 'Wood dust' gives a K_{st} of 208 bar m s^{-1} and a P_{max} of 9.4 bar. In the entire compilation for about 50 materials in [8] – coals and biomass – none is in an ST class higher than 2. Moving now to the work in the second row, the lower value for the char reflects the loss of volatiles during carbonisation. In reference [9] the K_{st} value for an anthracite is given as 1 bar m s^{-1}. This is due to the same effects, an anthracite being very low in volatiles. It will however also lack the pore structure of a brown coal char, which is why its K_{st} is so much lower than that for the char. The results in row 4 were obtained in a 20 litre vessel. From this and the $(dP/dt)_{max}$ value it is possible to calculate K_{st} as:

$$\mathbf{325\ bar\ s^{-1} \times (0.02)^{1/3}\ bar\ m\ s^{-1} = 88\ bar\ m\ s^{-1}}$$

a value almost equal to that for Yallourn coal. Moving to the fifth row, the LOC for a Spanish lignite is given as 8%, that for a German lignite as 10% and that for a Victorian brown coal as 13%. These of course have to be compared with the value of 21% for air. The term 'no development of combustion' in the table means that the kernel of suspended coal in air surrounding the scene of ignition does not respond to the ignition; in other words there is ignition but not propagation. In reference [13] (row 5) it is noted that a bituminous coal examined in the same way displayed a somewhat higher temperature, actually 1170–1350 K.

Table 17.1 Research investigations of dust explosions with lignites.

REFERENCE	DETAILS
[9]	The following information from cited literature given: K_{st} /bar m s^{-1} — P_{max} /bar Sapis, Italy lignite 162 6.8 Spanish lignite 107 8.8 German lignite 105 8.7 Morwell brown coal 220 7.6 Yallourn brown coal 91 6.7
[10]	Victorian brown coals Morwell: K_{st} 220 bar m s^{-1} Yallourn: K_{st} 91 bar m s^{-1} Brown coal char: K_{st} 64 bar m s^{-1}
[11]	Lignite (source not stated): K_{st} 196 bar m s^{-1}
[12]	Two lignites from Sulcis (see section 5.9), one from North Dakota. $(dP/dt)_{max}$ for the ND lignite ≈ 325 bar s^{-1}. For all three lignites P_{max} in the approximate range 6–6.5 bar
[13]	Limiting oxygen concentration (LOC), below which there is no development of combustion, given from independent literature
[14]	Spanish lignite. Methane present with the air at up to 6%. Temperature at the propagating front of a dust explosion between 950 and 1100 K

17.3.3 Case studies

An account of dust explosion hazards at the briquette works in Yallourn, Victoria is given in [15]. A 48-year-old man died at an explosion at the briquette works in Yallourn in 1953, and there were several non-fatal injuries [16]. Mining safety featured *en passant* in section 8.3 where the increased inherent safety of open cut mining over underground was referred to. Lignite from the underground mine at Dobrnja in Bosnia and Herzegovina was supplied to the power plant at Tuzla (see section 5.8.2). In 1990 there was an explosion at the mine which caused 180 deaths [17]. The initial explosion, which was caused by methane, occurred at a depth of about 450 m.

17.4 Spontaneous heating

There has been a huge amount of work on the spontaneous heating of lignites in storage and transportation, for example [18]. Reference [18] is from almost 25 years ago, but there is work long pre-dating it including that by the US Bureau of Mines. Here a few major points will be made. Low-rank coals are more susceptible than higher rank ones to spontaneous heating in stockpiles and shipping holds. Whenever a lump of brown coal is broken so that new surface is created, that exacerbates the spontaneous heating hazard. Fines in a stockpile area are a particular spontaneous heating hazard, another reason why breakage during stockpile assembly should be avoided. Water entering the pores in the internal structure of a lignite particle can, by releasing its heat of wetting, promote spontaneous heating. Incipient spontaneous heating can sometimes be detected by the human olfactory sense.

17.5 Concluding remarks

Seam fires, dust explosions and spontaneous heating have received attention in this chapter. Of course, the processes discussed previously such as hydrogenation have their particular hazards and are subject to the requirements of process safety including risk assessment and compliance with the requirements of regulatory bodies.

17.6 References

[1] http://madmikesamerica.com/2012/06/underground-coal-fire-in-the-mountain/

[2] http://mining.about.com/od/Coal/a/Minor-Coal-Fires-Hotspots-Bring-Major-Damage.htm

[3] Guhardja E. *Rainforest Ecosystems of East Kalimantan*, Springer Science & Business Media (2000)

[4] http://hazelwoodinquiry.vic.gov.au/

[5] Teague B., Catford J., Petering S. *Hazelwood Mine Fire Inquiry Report*, Victorian Government Printer (2014)

[6] Liu S., Wang C., Zhang S., Liang J., Chen F., Zhao K. 'Formation and distribution of polycyclic aromatic hydrocarbons (PAHs) derived from coal seam combustion: a case study of the Ulanqab lignite from Inner Mongolia, northern China', *International Journal of Coal Geology* 90: 126–134 (2012)

[7] http://www.tsmpacific.com/products/lc.htm

[8] http://www.explosiontesting.co.uk/explosion_ind_10.html

[9] Medina C.H., MacCoitir B., Sattar H., Slatter D.J.F., Phylaktou H.N., Andrews G.E., Gibbs B.M. 'Comparison of the explosion characteristics and flame speeds of pulverised coals and

biomass in the ISO standard 1 m^3 dust explosion equipment', *Fuel* 151: 91–101 (2015)

[10] Woskoboenko F. 'Explosibility of Victorian brown coal dust', *Fuel* 67: 1062–1068 (1988)

[11] Klippel A., Scheid M., Krause U. 'Investigations into the influence of dustiness on dust explosions', *Journal of Loss Prevention in the Process Industries* 26: 1616–1626 (2013)

[12] Continillo G., Crescitelli S., Furno E., Napolitano F., Russo G. 'Coal dust explosions in a spherical bomb', *Journal of Loss Prevention in the Process Industries* 4: 223–229 (1991)

[13] Mittal M. 'Limiting oxygen concentration for coal dusts for explosion hazard analysis and safety', *Journal of Loss Prevention in the Process Industries* 26: 1106–1112 (2013)

[14] Torrent J.G., Fuchs J.C., Borrajo J.L. 'On the combustion mechanism of coal dust in the presence of firedamp', *Combustion and Flame* 87: 371–374 (1991)

[15] Fletcher, M. *Digging People Up for Coal – A History of Yallourn*, Melbourne University Press (2002)

[16] http://www.virtualyallourn.com/node/27140

[17] http://www.nytimes.com/1990/08/27/world/178-miners-feared-dead-in-yugoslav-explosion.html

[18] Jones J.C., Vais M. 'Factors influencing the spontaneous heating of low-rank coals', *Journal of Hazardous Materials* 21: 203–212 (1991)

CHAPTER 18
LEONARDITE

18.1 Introduction

Leonardite has featured in sections 6.10 and 15.5.1 of this book. In this chapter its occurrence, nature and uses will be considered. The ability of leonardites to dissolve in alkali is their most notable feature. This is due to their humic acid content. Leonardite is so named after Dr A.G. Leonard, North Dakota's first State Geologist.

18.2 Sources of leonardite

A selection of such sources is given in Table 18.1.

Table 18.1 Sources of leonardite.

LOCATION OF DEPOSIT AND REFERENCE	DETAILS
Victoria, Australia [1]	400 million tonnes of leonardite at the mine at Maddingley
Saskatchewan [2]	Significant areas of leonardite within Estevan (see section 6.9.3)
North Dakota [3]	Leonardite between layers of clay at a deposit near Williston (see section 13.10.2). Major mining activity. Also leonardite mining at Scranton ND [4] and at Harmon (see section 15.5.3)
Greece [5]	Leonardite mines at Achlada and Zeli
Thailand [6]	See comments in the main text
Wyoming [7]	Major production at Converse Country
New Mexico [8]	Production at the San Luis mine
Germany	See comments in the main text
Spain [12]	Leonardite in the Maestrazgo basin

Leonardite is rich in humic acids as noted above, and consequently finds application as an organic fertiliser. Leonardite from Maddingley is used for this purpose. It is noted in reference [2] that leonardite is also believed to exist in British Columbia. The high quality of leonardite from the deposit near Williston is attributed to the layers of clay, which have prevented migration of contaminants into the leonardite layers. The leonardite from near Williston is used *inter alia* in drilling fluids. Wyoming also has leonardite reserves.

Reference [5] gives detailed information on the two Greek leonardites including petrographic analyses. In both the huminite group of macerals by far dominates, 96.7% for Achlada and 93.7% for Zeli (volume basis). They are significantly different in liptinite: Zeli 6.3% and Achlada 2.3%. Inertinite was measured in Achlada only as being 0.6%. The proximate analysis gave ash contents of 56% for Achlada and 58% for Zeli.

From the second and third rows of the table it is clear that leonardite and lignite often occur together, and this is true at Mae Moh in Thailand (see section 1.3.1 *et seq.*) where the leonardite is viewed as a waste. Mae Moh leonardite with and without carbonisation has been used on a trial basis as an adsorbent. In reference [7] attention is drawn to the fact that as a drilling fluid loss additive leonardite replaced quebracho, an extract from the Quebracho tree, which the US oil industry up to World War II imported from South America. It is pointed out in [9] that a substance to stain wood has been produced from New Mexico leonardite.

Leonardite from Germany is the basis of the manufacture of Powhumus®WSG 85, a plant growth stimulant obtained by alkali extraction from leonardite [10,11]. There are other 'commercial leonardites'. The point about uncertainty in rank made in section 5.5.1 and again in section 9.3 is relevant to the Maestrazgo basin (following row), coal from which has been called 'black lignite'. There are commercial leonardites originating in Spain [13].

18.3 Review of selected literature on leonardite

Table 18.2 gives details of some selected research studies of leonardites. Points of interest are brought out in the discussion which follows.

Table 18.2 Research investigations on leonardites.

REFERENCE	DETAILS
[14]	North Dakota leonardite studied alongside a lignite and a peat. Sufficient difference in humic acid composition between the lignite and the leonardite to provide the basis of a distinction between the two
[15]	Leonardite from North Dakota. Humic acid analysis alongside that of a lignite from the Maestrazgo basin (see Table 18.1)
[16]	Pyrolysis of a North Dakota leonardite to form an activated carbon
[17]	Three North Dakota lignites and two others examined by thermal analysis in air
[18]	Leonardite from Thailand, seen as a waste (see Table 18.1, row 3 and comments thereon). Carbonisation to make an adsorbent product

In the work in the first row, the peat was readily distinguishable from the lignite and the leonardite by its higher proportions of oxygen-containing functional groups in the humic acids, including carboxyl and phenol. Even so, as pointed out in the table the lignite and the leonardite could be distinguished from each other. The work in the second row has a similar theme: humic acid analyses of a lignite and a leonardite. Anomalously, in this work the humic acid from the leonardite was less oxidised than that from the lignite, though as the two were from different deposits this is probably insignificant. The leonardite humic acids contained more high molecular weight material than did those from the lignite. In the thermal analysis work (row 4) each lignite examined in the thermogravimetric unit displayed two peaks. Across the five, the first peak began at a temperature in the range 246–308°C and the second peak in the range 391–437°C. Overlap of devolatilisation and combustion, each with its kinetic scheme, is likely to be the origin of the behaviour whereby there are two maxima in the thermogravimetric trace. The following row is concerned with leonardite from Thailand, pyrolysed at temperatures up to 800°C to make an adsorbent material. The effect reported in section 12.3.2 for carbonisation of a Baori lignite was reported for the Thai leonardite: a lower surface area at the highest carbonisation temperatures. The leonardite chars in [18] had surface areas up to 66 $m^2 g^{-1}$.

18.4 Concluding remarks

Clearly the value of leonardites is in their humic acid content, which makes them suitable for their primary applications as organic fertilisers and in drilling fluids as loss additives. Distinction is not always made between leonardite deposits and upper layers of lignite deposits showing some oxidation, and whilst the former is a resource the latter is not generally seen as being although, as we saw in relation to the Mae Moh deposit in Thailand, it can be put to use.

18.5 References

[1] http://www.hiwtc.com/photo/products/10/00/13/1307.jpg

[2] Hoffman G.L., Nikols D.J., Stuhec S., Wilson R.A. *Evaluation of the Leonardite (Humalite) Resources of Alberta*, Open File Report 1993-18, Alberta Research Council

[3] http://www.leonarditeproducts.com/AboutUs.aspx

[4] *Minerals Yearbook 2007, Volume 2 Area Reports, Domestic*, US Department of the Interior

[5] Kalaitzidis S., Papazisimou S., Giannouli A., Bouzinos A., Christanis K. 'Preliminary comparative analyses of two Greek leonardites', *Fuel* 82: 859–861 (2003)

[6] Katanyoo S., Naksata W., Sooksamiti P., Thiansem S., Arquero O-R. 'Adsorption of zinc ions on leonardite prepared from coal waste', *Oriental Journal of Chemistry* 28: 373–378 (2012)

[7] *Mineral Occurrence and Development Potential Report*, US Department of the Interior, Bureau of Land Management Casper Field Office, Wyoming (2004)

[8] http://mines.findthedata.com/d/p/Leonardite

[9] Shomaker J.W., Hiss W.L. 'Humate mining in north western New Mexico', *New Mexico Geological Society 25th Field Conference*, Ghost Ranch, NM, pp. 333–336 (1974)

[10] http://www.humintech.com/001/agriculture/products/powhumus.html

[11] Danuta Sugier D., Kołodziej B., Bielińsk E. 'The effect of leonardite application on *Arnica montana* yielding and chosen chemical properties and enzymatic activity of the soil', *Journal of Geochemical Exploration* 129: 76–81 (2013)

[12] Olivella M.A., Gorchs R., de las Heras F.X.C. 'Origin and distribution of biomarkers in the sulphur rich Utrillas coal basin – Teruel mining district – Spain', *Organic Geochemistry* 37: 1727–1735 (2006)

[13] Cavani L., Ciavatta C., Gessa C. 'Identification of organic matter from peat, leonardite and lignite fertilisers using humification parameters and electrofocusing', *Bioresource Technology* 86: 45–52 (2003)

[14] Francioso O., Sanchez-Cortes S., Tugnolic V., Marzadoria C., Ciavatta C. Spectroscopic study (DRIFT SERS and 1H NMR) of peat, leonardite and lignite humic substances, *Journal of Molecular Structure* 565: 481–485 (2001)

[15] Garcia D., Cegarra J., Abad M. 'Comparison between alkaline and decomplexing reagents to extract humic acids from low rank coals', *Fuel Processing Technology* 48: 51–60 (1996)

[16] *Fuel and Energy Abstracts* September 1995, p. 335

[17] Francioso O., Montecchio D., Gioacchini P., Ciavatta C. 'Thermal analysis (TG–DTA) and isotopic characterization (^{13}C–^{15}N) of humic acids from different origins', *Applied Geochemistry* 20: 537–544 (2005)

[18] Chammui Y., Sooksamiti P., Naksata W., Thiansem S., Arqueropanyo O. 'Removal of arsenic from aqueous solution by adsorption on leonardite', *Chemical Engineering Journal* 240: 202–210 (2014)

CHAPTER 19

EXAMPLES OF CARBON CAPTURE AND STORAGE (CCS) AT LIGNITE-UTILISING PLANTS

19.1 Introduction

Carbon mitigation has featured in a number of previous sections of this book including 4.3. It was also noted that such reduction is one of the motivations for use of supercritical steam in power generation. This chapter is concerned with lignite-fired plants which use CCS or are developing it, and these are few and far between. None of the ND or TX plants covered in Chapter 6 carries out carbon capture at this time. Plans for CCS at Antelope Valley ND (see section 6.2.4) were cancelled [1] although plans for CCS at South Heart ND are on track (see Table 19.1).

19.2 Examples

Table 19.1 gives some examples of carbon capture at lignite-utilising enterprises.

Table 19.1 Carbon capture and storage at lignite-fired utilities.

LOCATION AND REFERENCE	DETAILS
Boundary Dam, Saskatchewan [2]	CO_2 capture by Shell Global Cansolv at one of the several power production units comprising the power plant
Quintana South Heart Project, North Dakota [4–6]	IGCC. Carbon capture expected to begin in 2018
Hazelwood, Victoria, Australia [7] (see section 10.2.2 et seq.)	Demonstration project. CO_2 capture by the locally developed UNO Mark 3 process
Schwarze Pumpe power plant, Germany (see section 4.3 et seq.) [8]	See comments in the main text

The entry in the first row gives details of carbon capture at Boundary Dam additional to those in section 6.9.2. Shell Global Cansolv involves reacting the carbon dioxide in the post-combustion gases with an amine [3], the chemistry being quite classical:

$$CO_2 + 2(R\text{-}NH_2) \rightarrow R\text{-}NHCOO^- + R\text{-}NH_3^+$$

and a modified chemical equation applies for a tertiary amine. The simplicity of the above however does less than justice to the Shell Global Cansolv approach, which as well as being applied in ways most suitable for particular situations enables the carbon dioxide to be regenerated and used in enhanced oil recovery (EOR) or, if that is not its intended destiny, to be collected in readiness for sequestration. At Boundary Dam, where capture began in October 2014, most of the 1 million tonnes of carbon dioxide captured over the first year will be directed at EOR. It is the world's first CCS on a full size (not pilot or demonstration scale) coal-fired plant. That it should be at a lignite-fired plant is in a way auspicious for the future of lignite utilisation.

At the Quintana South Heart Project (following row) the lignite fuel is gasified to make hydrogen according to:

$$C + 2H_2O \rightarrow CO_2 + 2H_2$$

and the carbon dioxide is removed at that stage. The hydrogen is burnt in an IGCC arrangement. At Quintana South Heart performance figures expected for 2018 are as follows [5]: 175 MW of electricity; hydrogen production at 4.7 million cubic metres per day; 2.1 million tonnes per annum of carbon dioxide. Simple calculations will be attempted for Quintana South Heart to see how the data given fit together. These are shown in the box.

4.7×10^6 m^3 of hydrogen per day will release, using a calorific value of 11 MJ m^{-3}:

$4.7 \times 10^6 \times 11 \times 10^6$ J $= 5.2 \times 10^{13}$ J per day

For round-the-clock operation this is a rate of heat supply of:

$[5.2 \times 10^{13}/(24 \times 3600)]$ W $= 600$ MW

Now carbon dioxide capture and removal have their energy requirements and this obviously reduces the overall efficiency of an IGCC process. Such energy has to be deducted from that which becomes the saleable product of the power plant. Details and practices vary, and so therefore do overall efficiencies. There is no generic value for the efficiency of an IGCC plant with CCS. What is clear is that the efficiency of an IGCC process without CCS is measurably higher than that of one with CCS, and that the difference depends *inter alia* on the electricity production rate. Now using the 175 MW of electricity reported for Quintana South Heart and the 600 MW of heat calculated above, the efficiency would be:

$(175/600) \times 100\% = 29\%$

and this, depending on the energy requirements of CCS as noted, is in the correct neighbourhood. The matter of losses through carbon dioxide capture is further examined in section 19.3.

Quintana South Heart is an example of pre-combustion capture of carbon dioxide. Its intended use is EOR for which pipelining will be needed. The compression which this requires has a not insignificant energy requirement.

At Hazelwood (next row) carbon dioxide removal is by contact with a solution of potassium carbonate, effecting conversion to the bicarbonate according to:

$$\mathbf{K_2CO_3(aq) + H_2O(l) + CO_2(g) \rightarrow 2KHCO_3(aq)}$$

Now the solubility of potassium carbonate in water at 40°C is 117 g per 100 g water. That of potassium bicarbonate at the same temperature is 47.5 g per 100 g water. In the chemical equation above, therefore, there is movement towards slurry conditions, and it is in the form of a slurry that the bicarbonate is concentrated by means of a cyclone. The carbon dioxide on the left-hand side of the chemical equation above has been very effectively 'captured' in what, as noted in the table, is called the UNO Mark 3 process. There is also activity at Hazelwood into pipeline transfer to offshore storage sites of carbon dioxide not having been so captured.

In moving to Schwarze Pumpe on the following row, we have to note that its operator Vattenfall have ceased all activity into carbon capture [9]. Capture at Schwarze Pumpe was to have involved oxy-fuel combustion [8]. In oxy-fuel combustion the fuel is supplied with oxygen, nitrogen having been removed from the oxidant supply. The post-combustion gas is therefore composed entirely of carbon dioxide plus water, the latter having of course originated from the lignite's own moisture content. The flue gas is therefore more concentrated in carbon dioxide than that from combustion with air, and separation of the water is very easy, relying only on elementary dew point principles. The carbon dioxide can then be stored for subsequent sequestration, use in EOR or whatever.

19.3 Electrical output penalty

The loss of output of power due to the requirements of CCS were examined above for an IGCC plant. Calculated results for conventional lignite-fired power station performance are given in [9]. Across six sets of conditions leading to power ranging from 527 to 569 MW (mid range value 548 MW) the daily capture of carbon dioxide was 5000 tonnes, representing about 35% capture. The electricity penalty for capture of one tonne was in the range 233–284 kW-hours, a mid range value 259 kW-hours. Using the mid range values, the proportion of the electricity generated required for capture is:

$$\mathbf{[5000 \times 0.259\ MW\text{-}hours/(548 \times 24\ MW\text{-}hours)] = 0.1\ (10\%)}$$

In many cases the penalty is considerably higher than this. Reference [10] states that 16% is common. This figure is for capture only: sequestration will obviously add to it. The 'electricity penalty of capture' does not equate to the electricity required for capture. Regeneration of the amine reagents requires steam, therefore there is less steam for generation which consequently drops [11].

19.4 Concluding remarks

Removal of carbon dioxide from flue gases is important as the world moves towards greater control of carbon dioxide emissions. Such techniques will be greatly helped by improved efficiencies of combustion, as already noted, and one wonders whether that is where there is really the most promise. The suspension of CCS research by Vattenfall was noted above. To this we can add that RWE Power has put on hold its development of CCS at Hurth [12]. To interpret this in broad terms is difficult, but from what has been said in this paragraph one might speculate that Germany, long a leader in matters relating to lignite utilisation, is heading for the 'increased combustion efficiency' approach to carbon mitigation. It is in any case clear that newly installed lignite-fired power plants in Germany will use supercritical steam.

19.5 References

[1] https://www.lignite.com/mines-plants/power-plants/antelope-valley-station/

[2] http://www.globalccsinstitute.com/project/boundary-dam-integrated-carbon-capture-and-sequestration-demonstration-project

[3] http://www.shell.com/global/products-services/solutions-for-businesses/globalsolutions/shell-cansolv/shell-cansolv-solutions/co2-capture.html

[4] http://www.globalccsinstitute.com/project/quintana-south-heart-project

[5] http://www.zeroco2.no/projects/south-heart-igcc

[6] http://www.sourcewatch.org/index.php/Quintana_South_Heart_Power_Project

[7] http://www.co2crc.com.au/research/uno_mk3_process.html

[8] http://spectrum.ieee.org/energywise/green-tech/clean-coal/vattenfall-ditches-carbon-capture-and-storage-research

[9] Nikolopoulos N., Violidakis I., Karampinis E., Agraniotis M., Bergins C., Grammelis P., Kakaras E. 'Report on comparison among current industrial scale lignite drying technologies (A critical review of current technologies)', *Fuel* 155: 86–114 (2015)

[10] Hammond G.P., Spargo J. 'The prospects for coal-fired power plants with carbon capture and storage: a UK perspective', *Energy Conversion and Management* 86: 476–489 (2014)

[11] http://www.rh.gatech.edu/news/60072/steam-process-removes-carbon-dioxide-regenerate-capture-materials

[12] http://www.zeroco2.no/projects/rwe-igcc-plant-with-co2-storage

CHAPTER 20
CO-COMBUSTION OF LIGNITES WITH OTHER FUELS

20.1 Introduction

There are many examples of co-combustion of lignites with other solid fuels. With carbon dioxide emission reduction such a pressing issue internationally, the most important is co-combustion of biomass with lignite. The basis of this was explained in section 4.2.4, where the co-firing of lignite with paper sludge at Frimmersdorf was described. In this chapter co-combustion is examined more widely.

20.2 Co-combustion with biomass

Table 20.1 gives details of some such enterprises, and comments on it follow below.

Table 20.1 Examples of lignite–biomass co-combustion.

LOCATION AND REFERENCE	DETAILS
Turkey [1]	Five lignites, including one from Tuncbilek and one from Elbistan, co-combusted with sunflower seed shells in a thermogravimetric analyser (TGA) at temperatures up to 900°C
Bosnia and Herzegovina [2]	Spruce sawdust co-combustion with lignites (one from Kakanj: see section 5.8.2)
Greece [3,4]	Kardia power plant (see section 5.3.6): proposal for co-firing lignite with cardoon (*Cynara cardunculus*), a thistle-like plant
South Australia [6]	The possibility of co-firing Leigh Creek coal (see section 10.3) with olive husks examined

In the work in row 1, the effect of the biomass on the lignite burnout yields was the point of greatest interest. Effects varied across the set of samples studied, in one case increasing the degree of burnout but for the other four decreasing it. The matter of 'unburnt carbon' in the ash from coal combustion is of course a highly classical one appertaining to such things as travelling grate stokers and their simulation in a pot furnace, and the results in [1] are relevant to that for whatever combustion plant is proposed for the co-combustion on a full scale. Two lignites were examined in the work described in the following row although one is classified as a brown coal. It was noted in sections 5.1.1 and 5.8.1 that the distinction persists respectively in the Czech Republic and in the former Yugoslavia. The lignite other than that from Kakanj was one used at Tuzla (see section 5.8.2). For the Karanj coal sulphur dioxide emissions were measured, and the well-known effect reported in section 5.3.3 for a Greek lignite – the trapping of sulphur dioxide as a solid sulphate by the calcium content of the lignite – was observed. This effect does not usually occur to the full extent possible, that is, the calcium is not quantitatively sulphated. With the Karanj coal it was found that the extent of sulphation was higher in the presence of the biomass, a favourable effect reducing sulphur dioxide emissions. The reason for this is probably the effect of the biomass on the combustion temperature.

Choice of biomass is one point of interest in the activity directed at the Kardia power plant (following row). When biomass is being used in electricity generation the regulatory bodies are entitled to know its origin and, if it involves the felling of trees, to satisfy themselves that new plantings are replacing whatever is harvested as a fuel [5]. Cardoon (*Cynara cardunculus)* for such use in Greece is being cultivated for the purpose, not reaped from the wild, but its occurrence is by no means restricted to Greece. Olive husks were the form of biomass considered for use with Leigh Creek coal (final row of the table). It is noted in [6] that olive husks are dense and, having naturally a low water content, have a calorific value about the same as that of the Leigh Creek coal. These two factors will minimise the need for plant modification in the event that co-firing is adopted. (See also section 11.3.)

20.3 Co-combustion with solid wastes

Research investigations into this include one involving Mae Moh lignite from Thailand, which has featured several times previously in this book. In [7] Mae Moh lignite is co-fired with municipal solid waste (MSW) in a test scale fluidised bed. MSW was present at up to 40% by weight. With increasing proportions of MSW the bed temperature decreases, whilst the temperature above the bed increases. Clearly this is due to devolatilisation of the MSW constituents within the bed and their burning above the bed. Components of the MSW displaying ready devolatilisation will include paper and cardboard, and quite possibly products above the bed will be accompanied by hydrocarbon gases originating from breakdown of the plastics. As recorded in section 7.4.2, Mae Moh lignite is high in sulphur. There is no reason why MSW should be, so one expects that co-combustion will reduce the sulphur dioxide yield. This was indeed so, reductions of up to 18% in the sulphur dioxide being observed when MSW was co-combusted with the lignite. This is a significant bonus.

Notwithstanding the total absence of coal-fired power plants in Nepal referred to in section 8.4, there has been interest there in making solid fuels by combining a lignite with combustible solid waste [8]. Briquettes were manufactured from blends of lignite and polythene, the latter simulating the plastic component of MSW. Polythene was present at up to 30% by weight, and briquetting pressure was 3.12 tons per square cm (≡ 275 MPa), and this can be compared with briquetting pressures given in section 11.2. The calorific value of the briquettes rose with polythene content. They were binderless and their mechanical durability was not high.

20.4 Co-combustion with higher rank coals

The co-firing of Texas lignite with sub-bituminous coal from Powder River at two power plants operated by Luminant was discussed in section 6.4.4. Co-firing of lignites and bituminous coals is described in [9], which uses p.f. in a furnace producing 15 kW of heat. The work uses an Indonesian lignite and bituminous coals from Indonesia and Canada and emphasis is on NO_x production. Co-combustion of bituminous coal with lignite leads to a drop in NO_x because of the reducing atmosphere created by the volatiles from the lignite. To the above should be added the point made in section 11.2.1 that higher ranks coals are sometimes used as binders in lignite briquettes.

20.5 Co-combustion with peat

This was examined in [10], where a Newfoundland peat and Canadian lignite were co-fired as a blend in a pilot scale furnace. The peat is much lower in ash than the lignite (see section 21.4) therefore co-firing with lignite did reduce the ash resulting from combustion but not in the simple way that might have been predicted from blend composition as other effects were operating. These include the formation of calcium chloride as a result of the unusually high (0.2%) chlorine content of the peat.

20.6 Concluding remarks

One can hope for developments in co-firing lignites with other fuels. The importance of co-firing with biomass has already been emphasised, and co-firing with MSW has the potential to expand. Several power plants co-firing MSW with higher rank coal are already up and running and supplying to the grid [11].

20.7 References

[1] Haykiri-Acma H., Yaman S. 'Effect of biomass on burnouts of Turkish lignites during co-firing', *Energy Conversion and Management* 50: 2422–2427 (2009)

[2] Kazagic A., Smajevic I. 'Synergy effects of co-firing wooden biomass with Bosnian coal', *Energy* 34: 699–707 (2009)

[3] Karampinis E., Nikolopoulos N., Nikolopoulos A., Grammelis P., Kakaras E. 'Numerical investigation Greek lignite/cardoon co-firing in a tangentially fired furnace', *Applied Energy* 97: 514–524 (2012)

[4] http://www.lignite.gr/abstracts/c_12GreekLignite.htm

[5] Jones J.C. *Global Trends and Patterns in Carbon Mitigation*, Ventus Publishing, Fredricksberg (2013)

[6] Dally B., Mullinger P. *Utilization of Olive Husks for Energy Generation: A Feasibility Study Final Report*, South Australian State Energy Research Advisory Committee (2002)

[7] Suksankraisorn K., Patumsawa S., Vallikul P., Fungtammasan B., Accary A. 'Co-combustion of municipal solid waste and Thai lignite in a fluidized bed', *Energy Conversion and Management* 45: 947–962 (2004)

[8] Shrestha A., Singh R.M. 'Energy recovery from municipal solid waste by briquetting process: evaluation of physical and combustion properties of the fuel', *Nepal Journal of Science and Technology* 12: 238–241 (2011)

[9] Moon C., Sung Y., Eom S., Choi G. 'NO$_x$ emissions and burnout characteristics of bituminous coal, lignite, and their blends in a pulverized coal-fired furnace', *Experimental Thermal and Fluid Science* 62: 99–108 (2015)

[10] Shao Y., Wang J., Xu C., Zhu J., Preto F., Tourigny G., Badour C., Li H. 'An experimental and modeling study of ash deposition behaviour for co-firing peat with lignite', *Applied Energy* 88: 2635–2640 (2011)

[11] Jones J.C. *Thermal Processing of Wastes*, Ventus Publishing, Fredricksberg (2010)

CHAPTER 21
COMPARISONS WITH PEAT

21.1 Introduction

In the earliest parts of the book it was explained how peat is the precursor to coal, and how in people's minds a similarity is naturally perceived between lignite and brown coal. This chapter will compare peat point by point with lignites. Again, research literature will be drawn on.

21.2 Petrographic comparison

Maceral formation is commonly seen as the result of organic metamorphism as explained in section 1.2. The maceral concept is however sometimes applied to peats, not arbitrarily but on the basis of observation. For example [1], an examination of a particular peat from Sumatra, revealed huminite macerals. These conformed in every respect to the criteria for classification. Literature pre-dating [1], for example [2,3], refers to the huminite content of peats. A peat from India when petrographically examined gave a vitrinite reflectance of 0.22 [4].

In [5] there is discussion of an immature brown coal and a peat which occur in the same basin in China. For 13 peat samples from this basin and for four brown coal samples, breakdown into maceral groups is given. For the peat samples the huminite content is in the range 75–98% and the reflectance in the range 0.10–0.19%. For the brown coal samples the huminite reflectances are in the range 0.28–0.33%. Distinction of a peat from a brown coal by huminite reflectance is therefore clear. The acceptability of this view is increased by the fact that in the work under discussion the peat and the brown coal are from the same basin, as noted.

The authors of [5] go on to define another maceral group which they call liptohuminite, distinguished from huminite on the basis of its higher reflectance and its colour when viewed in white light. There is an important further distinction when microfluorescence photometry is applied. The maceral sporinite[29], a member of the liptinite group, is capable of fluorescing when exposed to short-wavelength visible or u.v. light. That means that it emits light, and the fluorescence intensity decreases with rank. In [5] the fluorescence intensities for the liptohuminite macerals were higher than for the huminite macerals. Across the peat samples the percentage of liptohuminite ranged from 1 to 20. It was determined for only one of the brown coals, at 2%.

Sporinite is in the liptinite group of macerals. It has been noted [6] that in hydrogenation sporinite reacts very sluggishly in contrast to resinite. This has been widely observed and was commented upon in relation to lignite from Martin Lake TX (see section 6.4.5). The effect is exacerbated by the fact that Martin Lake lignite is 13% inertinite.

29 Sporinite features in schemes for petrographic classification of coals including reference [26] in Chapter 2.

In reference [7] peat and lignite from the same basin in Spain are examined, largely from the point of view of the oxygenated organic compounds they contain. There was found to be no non-arbitrary demarcation between the peat and the lignite. This is expressed in [7] by the term 'gradient in maturity'.

There is a great deal of subsea peat. One place where it has been gathered in sample quantities and examined [8] is the North Sea, off the Shetland Islands, also a major scene of oil and gas production.

The timescale of formation of peat reserves is millennia (e.g. [9]) in contrast to that of the formation of lignites, which is expressed in mega years (My or Ma) [10]. As examples, Bełchatów lignite belongs to the Miocene epoch [11], which was from 5.3 to 23 million years ago [12]. Louisiana lignite is more mature, belonging to an epoch bracketed by time from 36 to 66 Ma [13]. Either of these is many orders of magnitude higher than the time required for peat formation. Lignites from a Miocene deposit in Kalimantan, Indonesia when examined petrographically [14] were deemed to have advanced along the peat-to-lignite sequence only to a partial degree. The lignite in Tuscany referred to in section 5.9 is dated 8.5–9 Ma, which classifies it as Miocene. A lignite has never 'arrived': at most it has coalified to a degree where its distinction from sub-bituminous becomes unclear. The uncertainty in classification on the basis of age has been noted in relation to Beauchêne Island, one of the Falklands group, where a 'lignitic' deposit is, at an estimated age of 1250 years, declared to be 'several hundred times too young to be a true lignite' [15]. It was noted in Chapter 3 that the Clovis people had used lignite about 15 000 years ago. This is about the timescale for peat deposit formation. The lignite at Legler, New Jersey also co-exists with peat [16] and has sometimes not been distinguished from the peat in descriptions of the deposit. The lignite shows variations in colour, probably not a genuine lithotype effect but due to high mineral contents. It was observed that layers of clay above or below those of lignite contain some plant debris. Peat and lignite co-exist at many other places including Hula, Israel (section 7.9).

21.3 Calorific values and combustion

As is clear from Table 1.1, the calorific value of a lignite depends on the extent of moisture loss from the initial condition in the bed. The same is true of peat, which in the bed-moist state can have a moisture content of 90%. Table 20.1 gives some reported calorific values for peats.

Table 21.1 Calorific value of selected peats.

ORIGIN OF THE PEAT AND REFERENCE	CONDITION OF THE PEAT AND ITS CALORIFIC VALUE
Canada [17]	Fully dried: 22.5 MJ kg^{-1}
Finland [18]	Fully dried: 23.8 MJ kg^{-1}
Canada [19]	Moisture content 25%: 15.9 MJ kg^{-1}
Ireland [20]	Moisture content 25%: 16.3 MJ kg^{-1}
Indiana, USA [21]	Dry: 20.9 MJ kg^{-1}

The calorific values are of course relevant to power generation. Power generation with peat occurs, for example, in Ireland [22] where fluidised beds are used for the combustion. There has been electricity generation from peat in Finland since the 1970s, when it replaced imported fuel. For example [23], at Ilomantsi in Finland, near the border with Russia, there is electricity generation at 3.5 MW from the co-firing of peat with wood chips in a CHP arrangement. Fuel pellets comprising peat and wood waste are also made there. Internationally there is significant peat briquette production [24].

Peat, like lignite, is usually burnt in pulverised form when used to raise electricity (although, of course, fluidised beds can be used with either). A hammer mill can be used to pulverise peat [25]. Loesche (see sections 4.2.8 and 4.3), a major supplier of mills for pulverising lignite in Germany, also manufacture a mill which can be used to grind peat [26].

21.4 Ash-forming constituents

It is clear from Table 1.1 as well as from certain subsequent parts of this book that lignites can be very high in ash. That from Mae Moh in Thailand is an obvious example. In considering the ash contents of peats vis-à-vis those of lignites, it has to be remembered that the inorganic substances in peat are only those from the original plant deposition [27]. In coals of any rank, that can have been added to over geological time by marine incursion or by volcanic debris. Proximate analysis figures for peat are available in the literature. One such set of figures gives an ash content of 4.17% for a Russian peat in air-dried condition [28]. Another gives 3% for a Canadian peat [29]. In reference [27] it is noted that, in electricity generation using peat, slag formation can be sufficient to necessitate shut-down. Section 5.3.6 can be consulted for a comparison with such behaviour with lignites. Briquetting of peat is widely practised.

21.5 Carbonisation and gasification potential

Activated carbons from peat are commonplace. Peat can of course be used to make synthesis gas which can then be converted to liquid by F-T. The view that liquid fuels from peat via synthesis gas and F-T is likely to be viable in Finland has been expressed [30].

A Finnish peat, a Rhenish lignite and pine sawdust were examined for gasification performance in a fluidised bed in work reported in [31]. Gasification was in air and steam at temperatures up to 1000°C. Peat and biomass are the only locally available fuels in Finland. Depending on the proportion of air and the reacting pressure, carbon conversions with the peat were up to 95%. They were about the same as this for the sawdust and the lignite. Yields of gas were up to 3.5 m^3 per kg of peat, 2.5 m^3 per kg of sawdust and 4 m^3 per kg of lignite. There were significant amounts of methane and of C_2 hydrocarbons in the gaseous products, which obviously had the effect of raising their calorific values. With all three starting materials there was tar production, as high as 15% for the peat at the lower end of the temperature range; the tar yield with lignite was lower though not insignificant, and there is a basis for comparison with the liquid products from Victorian brown coal gasification described in section 13.2.

21.6 Conversion to liquid fuels

The suitability of peat for conversion to liquid with carbon monoxide is discussed in [32]. This corresponds to conversion of lignite to liquid with carbon monoxide and water, discussed in section 14.5.3; the peat as examined experimentally in [32] was 90% moisture, so this provided the water reagent. The conclusion was expressed that the most important factor in determining the suitability of a particular peat for conversion by this means is calorific value. Hydrogenation has also been looked into. For example, in [33] a Canadian peat treated with hydrogen gas gave 76% conversion to liquids with asphaltenes and oil in the products (cf. Table 14.2).

It is clear from the preceding two sections, brief though they are, that there has been interest in these methods for peats concurrently with their interest in relation to lignites. Production of chemicals follows.

21.7 Wax from peat

Table 21.2 gives details of some investigations into obtaining wax from peat.

Table 21.2 Examples of wax production from peat.

ORIGIN OF THE PEAT AND REFERENCE	DETAILS
Minnesota, USA [34]	Solvent extraction of wax from air-dried peat. Across a set of five peats wax yields ≈ 3%
Finland [35]	Wax deposition during drying at a peat-fired power plant
Finland [36]	Wax yields of 35–60 kg per tonne of peat

In [35] it was found that the melting points of the waxes had a dependence on the solvent used to extract them meaning, of course, that the composition also had such a dependence. There is continuity between the contents of the second row of the table and the mention earlier in the chapter of electricity production from peat in Finland. Oxygenated compounds – carboxylic acids, phenols and alcohols – were found to dominate in the wax composition.

21.8 Supercritical extractions

Work from about 30 years ago [37] describes supercritical extraction of chemicals from peat sourced in Minnesota. The fluids were water, methanol–water mixtures and acetone–water mixtures. Yields were in the range 19–50%. When peat originating in Canada was treated with supercritical water the products were oil and gas, distribution depending on whether a catalyst was or was not used [38]. A product identified as being particularly useful was the gas obtained when the process was catalytic, which had a calorific value of 27.7 MJ kg^{-1}. The gas produced under non-catalytic conditions had a calorific value on a weight basis only about a tenth of this.

21.9 Hazards with peat

Peat can undergo spontaneous heating, in which case it displays the same phenomenology that lignites do [39]. Fires frequently occur in beds of peat [40]. The propensity of dried peat to dust explosions has been noted in Finland [41].

21.10 White peat

The term possibly originates from the fact that such peat is to be found close to the White Sea off Russia [42]. White peat is applied to horticulture [43] and is less advanced in conversion from the original plant deposition than most peats.

21.11 Sapropels

The precursor to a sapropelic lignite, several of which have featured in earlier parts of the book, is not peat but sapropel (Greek: *sapros* 'decayed', *pelos* 'mud'). This is formed from aquatic plants and from other freshwater life including plankton. One respect in which sapropels differ from peats is their lower humic acid content [44]. In Belarus the reserves of 3 billion cubic metres of sapropel are seen as an important resource [45]. There are large amounts in Russia, where it is used as a fertiliser as it is in many other countries including the UK. The timescale of formation of sapropels is the same as that for peats (section 21.2). For example, a range of sapropels from beneath the Black Sea have been judged to be between 3000 and 7000 years old [46]. The calorific value of a sapropel from Lithuania when dried has been reported as being 20.6 MJ kg^{-1} [47].

21.12 Concluding remarks

With reference to the above two sections it is clear again that what has been done for lignites has, not necessarily to the same degree, been done for peat. That is the conclusion a reader studying this chapter will reach. Yet another similarity is that lignites and peats are both used in drilling fluids [48]. Sapropels have also been so used [49].

21.13 References

[1] http://specialpapers.gsapubs.org/content/286/63.abstract

[2] Vačeva S.P. 'Reflectance of macerals from bright brown coal, Pernik basin', *Fuel* 58: 55–58 (1979)

[3] McCartney J.T., Teichmüller M. 'Classification of coals according to degree of coalification by reflectance of the vitrinite component', *Fuel* 51: 64–68 (1972)

[4] Mukherjee A.K., Alam M.M., Ghose S. 'Microhardness characteristics of Indian coal and lignite', *Fuel* 68: 670–673 (1989)

[5] Kuili J., Yong Q. 'Coal petrology and anomalous coalification of Middle and Late Pleistocene peat and soft brown coal from the Tengchong Basin, Western Yunnan, People's Republic of China', *International Journal of Coal Geology* 13: 143–170 (1989)

[6] Cronauer D.C., Joseph J.T., Davis A., Quick J.C., Luckie P.T. 'The beneficiation of Martin Lake Texas lignite', *Fuel* 71: 65–73 (1992)

[7] del Rio J.C., Gonzalez-Vila F.J., Martin F. 'Variation in the content and distribution of biomarkers in two closely situated peat and lignite deposits', *Organic Geochemistry* 18: 67–78 (1992)

[8] Hoppe G. 'Submarine peat in the Shetland Islands', *Geografiska Annaler Series A, Physical Geography* 47: 195–203 (1965)

[9] *Towards an Assessment of the State of UK Peatlands*, Report No. 445, Joint Nature Conservation Committee, Peterborough (2011)

[10] http://www.geosociety.org/TimeUnits/viewComments.asp

[11] Drobniak A., Mastalerz M. 'Chemical evolution of Miocene wood: example from the Belchatow brown coal deposit, central Poland', *International Journal of Coal Geology* 66: 157–178 (2006)

[12] Levin H.L. *The Earth Through Time*, 10th edition, John Wiley (2013)

[13] *Lignite Resources in Louisiana*, Public Information Series No. 5, Louisiana Geological Survey (2006)

[14] Dwiantoro M., Notosiswoyo S., Anggayana K., Widayat A. 'Paleoenvironmental interpretation based on lithotype and macerals variation from Ritan's lignite, Upper Kutai Basin, East Kalimantan', *Procedia Earth and Planetary Science* 6: 155–162 (2013)

[15] http://www.nature.com/nature/journal/v309/n5969/abs/309617a0.html

[16] Rachelle L.D. 'Palynology of the Legler lignite: a deposit in the tertiary Cohansey formation of New Jersey, USA', *Review of Palaeobotany and Palynology* 22: 225–252 (1976)

[17] Oren M.J., MacKay G.D.M. 'Peat–water–oil mixture as a low cost liquid fuel', *Fuel* 69: 1326–1327 (1980)

[18] Björnbom E., Olsson B., Karlsson O. 'Thermochemical refining of raw peat prior to liquefaction', *Fuel* 65: 1051–1056 (1986)

[19] Haanel B.F. *Final Report of the Peat Committee*, Governments of the Dominion of Canada and the Province of Ontario (n.d.)

[20] O'Donnell S., *New Scientist* 4 July 1974, pp. 18–19

[21] Lyons R.E., Carpenter C.C. 'A chemical examination of and calorimetric test of Indiana peats', *Journal of the American Chemical Society* 30: 1307–1311 (1908)

[22] Sarkki J., Griffin F., Scully S., Flynn T. 'CFB technology in ESB peat burning power stations', *21st International Conference on Fluidized Bed Combustion* (2012)

[23] *CHP and Pellet Factory in Ilomantsi*, Infocard 5, Northern Wood Heat (n.d.)

[24] http://www.peatsociety.org/peatlands-and-peat/global-peat-resources-country

[25] Wahlstrom F., Kortela U. 'Combustion stabilisation and improvement of the efficiency in a peat power plant', *Real Time Digital Control Applications* (A. Alonso-Concheiro, ed.) pp. 173–182 (1983)

[26] http://en.siringos.com.mx/crushingplant/loesche-coal-grinding-roller-mill.php

[27] Heikkinen R., Laitinen R.S., Patrikainen T., Tiainen M., Virtanen M. 'Slagging tendency of peat ash', *Fuel Processing Technology* 56: 69–80 (1998)

[28] Kim J.W., Lee H.D., Kim H.S., Park H.Y., Kim S.C. 'Combustion possibility of low rank Russian peat as a blended fuel of pulverized coal fired power plant', *Journal of Industrial and Engineering Chemistry* 20: 1752–1760 (2014)

[29] Shao Y., Xu C., Zhu J., Preto F., Wang J., Tourigny G., Badour C., Li H. 'Ash and chlorine deposition during co-combustion of lignite and a chlorine-rich Canadian peat in a fluidized bed – effects of blending ratio, moisture content and sulfur addition', *Fuel* 95: 25–34 (2012)

[30] Kirkinen J., Soimakallio S., Makinen T., Savolainen I. 'Greenhouse impact assessment of peat-based Fischer–Tropsch diesel life-cycle', *Energy Policy* 38: 301–311 (2010)

[31] Kurkela E., Stahlberg P. 'Air gasification of peat, wood and brown coal in a pressurized fluidized-bed reactor. I. Carbon conversion, gas yields and tar formation', *Fuel Processing Technology* 31: 1–21 (1992)

[32] Björnbom E., Björnbom P. 'Some criteria for the selection of peat as a raw material for liquefaction', *Fuel* 67: 1589–1591 (1988)

[33] Cavalier J-C., Chornet E. 'Fractionation of peat-derived bitumen into oil and asphaltenes', *Fuel* 57: 304–308 (1978)

[34] Spigarelli S.A., Chang F.H., Kumari D. 'Bitumen and wax yields from wet-carbonized, hydrolyzed and untreated peats', *International Journal of Coal Geology* 8: 123–133 (1987)

[35] Fagernas L., Sipil K. 'The behaviour of waxy compounds in the drying system of a peat power plant', *Fuel Processing Technology* 21: 189–200 (1989)

[36] Spedding J. 'Peat' (Review), *Fuel* 67: 883–900 (1988)

[37] Scarrah W.P., Scarrah P., Myklebust L. 'The supercritical fluid extraction of peat using water and aqueous organic solutions', *Fuel* 65: 274–276 (1986)

[38] Xu C., Donald J. 'Upgrading peat to gas and liquid fuels in supercritical water with catalyst', *Fuel* 102: 16–25 (2012)

[39] Jones J.C. 'The oxidation of peat and its thermal accompaniment', *Journal of Chemical Technology and Biotechnology* 45: 223–229 (1989)

[40] http://www.environxsolutions.com/about-peat-fires

[41] Eckhoff R. *Dust Explosions in the Process Industries: Identification, Assessment and Control of Dust Hazards*, Gulf Professional Publishing (2003)

[42] Filatov N., Pozdnyakov D., Johannessen O.M., Pettersson L.H., Bobylev L.P. *White Sea: Its Marine Environment and Ecosystem Dynamics Influenced by Global Change*, Springer (2007)

[43] http://www.substrate-consulting.com/substrates/peat/

[44] Sokolov G., Szajdak L., Simakina I. 'Changes in the structure of the nitrogen-containing compounds of peat-, sapropel- and brown coal-based organic fertilisers', *Agronomy Research* 61: 149–160 (2008)

[45] http://www.sectsco.org/EN123/Belarus.asp

[46] Brown S.D., Chiavari G., Ediger V., Fabbri D., Gaines A.F., Galletti G., Karayigit A.I., Love G.D., Snape C.E., Sirkecioglu O., Toprak S. 'Black Sea sapropels: relationship to kerogens and fossil fuel precursors', *Fuel* 79: 1725–1742 (2000)

[47] Kozlovska J., Valančius K., Petraitis E. 'Sapropel use as a Biofuel Feasibility Studies', *Research Journal of Chemical Sciences* 2: 29–34 (2012)

[48] http://www.ogj.com/articles/print/volume-107/issue-9/drilling-production/special-report-new-olefin-based-drilling-fluid-improves-operational-environmental-profile.html

[49] Lishtvan I.I., Lozhenitsyna V.I., Lerman A.S., Lobov A.I., Artamonov V.Y. 'Sapropel drilling-mud infiltration into a porous medium', *Journal of Engineering Physics* 55: 1123–1127 (1988)

CHAPTER 22
COMPARISON WITH SUB-BITUMINOUS COALS

22.1 Introduction

The previous chapter was concerned with peat, the substance preceding lignite in the coalification sequence. It was shown how the two are sometimes indistinct, and that this is so at the boundary between lignite and sub-bituminous coal has already been mentioned, for example in sections 5.4.3 and 5.5.1. The co-firing of a lignite with sub-bituminous coal from Powder River Basin was described in section 6.4.4.

22.2 Petrographic composition

Sub-bituminous coals are more mature than lignites. Table 22.1 gives details of some sub-bituminous coal deposits, with such information as vitrinite reflectance.

Table 22.1 Details of selected sub-bituminous coals.

LOCATION	DETAILS
Collie, Western Australia	Permian (251–299 Ma). Vitrinite reflectance 0.32–0.47% [1]. Maceral analysis: vitrinite 37–57%; liptinite 8–14%; inertinite 23–48% [1]. A different source [2] gives a wider range of vitrinite contents: 26–79%. Calorific value as received 20 MJ kg^{-1}
Oaklands, New South Wales	Permian. Vitrinite reflectance 0.40 [4]
Talcher, India	Permian. Vitrinite reflectances ≈ 0.45 [5]. Significant sporinite (see section 21.2)
Central Alberta, Canada	Classifications into lithotypes: 'dull', 'banded dull', 'banded', 'banded bright', 'bright' [6]
Waikato, New Zealand	Sub-bituminous coal having vitrinite reflectance 0.4 [7]
Maryvale, New Zealand	See comments in the main text

Information on Collie coal in the first row establishes its higher rank than lignite. The Hardgrove indices of a number of Collie coals and blends thereof have been measured [3] as being in the range ≈ 48 to 58 when air dried. Coal from Collie is used at two power stations in Western Australia, with combined capacity approximately 1500 MW. The same contrast with lignites is evident in the case of Oaklands coal in the following row. Talcher coal (row 3) has been identified as being suitable to make synthesis gas for ammonia production. Lithotype, and its variation with depth, is used as a basis for classification with the Canadian coal in the next row, a common factor with lignites.

The sub-bituminous coal from Maryvale, New Zealand [8] is particularly high in a particular maceral of the huminite group, detrogelinite. This is in the range 49.8–53.5% of the total macerals across a range of four Maryvale coals studied. The term 'detrogelinite' does not feature in reference [26] in Chapter 2, which has been used as a source for such terminology previously in the book. One has to consider the nature of macerals generally in clarifying this, in particular the fact that maceral composition changes as a deposit advances along in degree of coalification. The precursor to detrogelinite is attrinite/densinite [9] and this does feature in the source previously drawn on. The arbitrariness of terminology in coal petrography has to be justified by two factors. First, a maceral, unlike its inorganic counterpart a mineral, does not have a precise composition, a fact to which the present author has drawn attention previously [10]. Secondly, maceral analysis is by microscopic examination and differences in the appearance of a particular maceral across ranks of coal might justify a name change. The background of a maceral over geological time is for many purposes less important than its accurate description in the present. It is this factor which has led to the differing formulations of gelification index which have been given in this book.

It sometimes happens at a deposit that there is a boundary between sub-bituminous coal and lignite [11]. The best example of this is the interface ('merger') of the Powder River Basin with the Fort Union area lignite deposit in Montana [12], for a cartographic representation of which a reader should go to [13]. Coal from Rosebud County Montana originates at the Powder River Basin. It has a nameplate capacity of 41.5 MW [14].

Examination of the Energy Information Administration map of 'Coal-bearing areas of the United States'[30] reveals that a lignite deposit in Colorado is encircled by sub-bituminous coal, the latter forming an annulus around the former. This is in the Denver basin (see section 13.2). The same map shows a major deposit of sub-bituminous coal in New Mexico where the boundary with bituminous coal almost coincides with the state boundary with Colorado. Similarly, at the large sub-bituminous coal reserve at Kuznetsk in western Siberia there is co-existence with bituminous coal [15].

This section then has related petrographic details for typical sub-bituminous coals to those for lignites. Carbonisation and gasification follow.

22.3 Carbonisation and gasification

Sub-bituminous coals when carbonised produce a char, not a coke. The distinction is of course that a coke is hard and fused, and is made by carbonising a suitable bituminous coal. By 'suitable' is meant one with a high coking propensity, which by no means all bituminous coals have. Activated carbons can be made from sub-bituminous coal. There have been investigations of UGC at sub-bituminous deposits including that at Hoe Creek, Wyoming [16]. The gas so manufactured was of composition hydrogen 37.5%, CO 23.5%, methane 5.5%, balance non-combustibles. It is left as an exercise for the reader to show that such a gas would have a calorific value of 9 MJ m^{-3}.

30 Accessible online.

22.4 Coal-bed methane from sub-bituminous coals

There is abundant methane at Powder River [17]. The same is true of some other deposits of sub-bituminous coal including that at Waikato (see Table 22.1) [18], where the methane content is typically 2.5 m^3 $tonne^{-1}$ of coal.

22.5 Conversion to liquid fuels

A Wyoming sub-bituminous coal examined in gaseous hydrogen at 6 MPa in a nickel–molybdenum catalyst gave a conversion of over 60% [19]. This work has been chosen as the basis of this section as a brown coal from Loy Yang was treated in the same way for comparison, and showed a conversion of up to 80%. Conversions were lower for both when a cobalt–molybdenum catalyst was used.

22.6 Humalite

This is the analogue for sub-bituminous coal of leonardite for lignites, being sub-bituminous coal partly oxidised. It occurs in Alberta [20] whence the name is derived: *Hum*(ic)*Al*(berta)-*ite*. There is considerable trade in it as a soil additive. Like lignites and peats, it is used in the production of drilling fluid additives [21].

22.7 Amounts of lignite and of sub-bituminous coal mined

In the USA sub-bituminous coal well exceeds lignite in production, 47% of the total tonnage of coal as compared to 7% for lignite [22]. The balance is bituminous with a very small (0.2%) amount of anthracite. Powder River supplies 90% of the sub-bituminous coal in the USA [23].

The differential in US production of lignite and sub-bituminous coal to some degree reflects world production of the two ranks of coal. It was reported in Chapter 1 that lignite accounts for 23% of the world's coal production on a tonnage basis. Sub-bituminous accounts for about 30% [24].

22.8 Hazards with sub-bituminous coals

They can display spontaneous heating, showing the same criticality behaviour that lignites and peat do. They can cause dust explosions, and ignition in layers. In each of these, lignites show greater reactivity attributable to their higher volatile content.

22.9 Concluding remarks

Continuity of geology between the two ranks of coal has been emphasised in this and previous chapters, and in summing up it is sufficient to comment that this is reflected in continuity of properties.

22.10 References

[1] Mishra H.K. 'Comparative petrological analysis between the Permian coals of India and Western Australia: paleoenvironments and thermal history', *Palaeogeography, Palaeoclimatology, Palaeoecology* 125: 199–216 (1996)

[2] Kershaw J.R., Taylor G.H. 'Properties of Gondwana coals with emphasis on the Permian coals of Australia and South Africa', *Fuel Processing Technology* 31: 127–168 (1992)

[3] Vuthaluru H.B., Brooke R.J., Zhang D.K., Yan H.M. 'Effects of moisture and coal blending on Hardgrove Grindability Index of Western Australian coal', *Fuel Processing Technology* 81: 67–76 (2003)

[4] Shiboaka M. 'Genesis of micrinite in some Australian coals', *Fuel* 62: 639–644 (1983)

[5] Mishra H.K., Chandra T.K., Verma R.P. 'Petrology of some Permian coals of India', *International Journal of Coal Geology* 16: 47–71 (1990)

[6] Demchuk T.D. 'Epigenetic pyrite in a low-sulphur, subbituminous coal from the central Alberta Plains', *International Journal of Coal Geology* 21: 187–196 (1992)

[7] Glombitza C., Mangelsdorf K., Horsfield B. 'Structural insights from boron tribromide ether cleavage into lignites and low maturity coals from the New Zealand Coal Band', *Organic Geochemistry* 42: 228–236 (2011)

[8] Crosdale P.J. 'Coal maceral ratios as indicators of environment of deposition: do they work for ombrogenous mires? An example from the Miocene of New Zealand', *Organic Geochemistry* 20: 797–809 (1993)

[9] Diessel C.F.K. *Coal-Bearing Depositional Systems*, Springer (2012)

[10] Jones J.C. 'The nature of macerals', *Fuel* 89: 1743 (2010)

[11] Thomas L. *Coal Geology*, John Wiley (2002)

[12] *Final Environmental Impact Statement: Proposed Federal Coal Leasing Program*, US Department of the Interior (1975)

[13] http://pubs.usgs.gov/ha/ha730/ch_i/gif/I054.GIF

[14] http://www.sourcewatch.org/index.php/Rosebud_Power_Plant

[15] Kuztneztsov P.N., Ilyushechkin A.Y. 'Coal resources, production and use in the Russian Federation', *The Coal Handbook: Towards Cleaner Production* (D. Osborne, ed.) Chapter 7 pp. 148–168, Elsevier (2013)

[16] Cooper B. *The Science and Technology of Coal and Coal Utilization*, Springer (2013)

[17] http://www.powderriverbasin.org/coalbed-methane/

[18] Zarrouk S.J., Moore T.A. 'Preliminary reservoir model of enhanced coalbed methane (ECBM) in a subbituminous coal seam, Huntly Coalfield, New Zealand', *International Journal of Coal Geology* 77: 153–161 (2009)

[19] Redlich P.J., Hulstona C.K.J., Jackson W.R., Larkins F.P., Marshall M. 'Hydrogenation of sub-bituminous and bituminous coals pre-treated with water-soluble nickel–molybdenum or cobalt–molybdenum catalysts', *Fuel* 78: 83–88 (1999)

[20] http://canadianhumaliteinternational.com/

[21] http://www.blackearth.com/downhole/

[22] http://www.eia.gov/todayinenergy/detail.cfm?id=2670

[23] http://what-when-how.com/energy-engineering/coal-supply-in-the-u-s-energy-engineering/

[24] *International Energy Outlook 2013 with Projections to 2040*, United States Energy Information Administration (2013)

Note added in proof: Readers might benefit from consulting *'Sub-bituminous coals: An Overview'* by J.C. Jones, published as an e-book by Ventus in 2015. Downloading is free.

CHAPTER 23
LIGNITE ORIGINATING IN ISOLATED OR UNDEVELOPED LOCATIONS

23.1 Introduction

There are other countries and locations meriting mention whose inclusion in the mainstream discussion was precluded by geographical location and isolation from major centres of activity in lignite utilisation. Such regions are discussed in this chapter, as well as a few others where lignite discovery was an almost incidental result of a geological survey, prospects of utilisation being extremely small. This point was touched on in section 6.11, where the lignite deposit at Ocean Beach in San Francisco was mentioned.

23.2 Examples

Examples are shown in Table 23.1, and comments follow below.

Table 23.1 Lignite originating in isolated or undeveloped locations.

	COUNTRY OR REGION	DETAILS
1	South Africa	Significant reserves, not mined since the early 1990s [1]. The three major deposits – those at Koekenaap, at Bergrivier and at Kraaifontein – are in Cape Province [2]
2	Ethiopia	Major deposits of lignite, and plans to place a power station close to one of them [3]
3	Trinidad and Tobago	Lignite discovered in Trinidad in the mid 19th century [4]. Never commercially developed
4	Falkland Islands	See section 21.2
5	Madagascar	Significant lignite [5]
6	Malawi	Lignite (about 2 million tonnes) to the exclusion of higher rank coals [6]
7	Guam (US territory)	Lignite in small quantities [9]. No utilisation
8	Eritrea	Some lignite, only ever mined on a very small scale [10]
9	Jamaica	Some lignite, used for heating [11]
10	Eastern Arctic Archipelago (Canada)	Lignite in addition to higher rank coals. 15 000 million tonnes of lignite in the area surveyed in [13]

Cont'd...

Table 23.1 Cont'd...

11	Kerguelen Islands, French Southern and Antarctic Lands	Lignite known to exist [14]
12	Jammu and Kashmir	See comments in the main text
13	West of Mertz Glacier, Antarctica	Lignite discovered during a geological survey [17]
14	Australian interior	Lignite known to exist in the Lake Eyre Basin [18]
15	Afghanistan	Lignite discovered near Zurmat in the 19th century [19]
16	Bhutan	See comments in the main text
17	Hudson Bay Lowland	Lignite, noted in [22] to have been not yet utilised
18	Papua New Guinea (PNG)	Lignite in north-eastern PNG [23] and in the west [24]
19	Iceland	Lignite at the Tjörnes Peninsula in northern Iceland [25]. Lignite once mined in very small amounts in a different part of Iceland [26]
20	Dominican Republic	Approximately 70 million tonnes of lignite [28]. Also lignite in neighbouring Haiti
21	Off Barbados	'Lignite debris' in a mineral formation [30]
22	Alaska	See comments in the main text
23	Nicaragua, Atlantic coast	Lignite found in exploration wells for oil and gas [34]

With reference to South Africa, it is interesting to note that stoppage of production coincided with the dismantling of Apartheid with all that that meant in terms of international trade. The lignites in South Africa are 'lenticular' [22], and consistently with what has been said earlier there is evidence of their being allochthonous. Even so there are roots evident, and this points to an autochthonous contribution.

Ethiopia (population 94 million) at present has 792 MW of installed power almost all of it hydroelectric. The proposed power plant, to be known as the Yayu Coal Mine and Thermal Power Plant, will be at Achibo which is about 340 miles from Addis Ababa. Its location was obviously determined by that of the lignite reserve. It will contribute 100 MW which, as well as boosting the national capacity by 12.5%, will help ease the almost total dependence on hydro. The project has stalled more than once over the years since its inception through lack of government backing, partly caused by awareness that being lignite-fired the plant would add to carbon dioxide emissions.

Madagascar (row 5) is an example of a country where development of utilisation has been prevented by lack of funds. The lignite in Malawi would be sufficient to produce 2 TW-hour of electricity. This is almost exactly equal to the annual electricity production of the country [7] most of which is hydroelectric [8]. This is also by coincidence the annual production in Guam (following row) where generation is entirely thermal with imported fuels. In [10] it is noted that the lignite in Eritrea was deemed unsuitable for steam locomotive use. This mirrors experience in Victoria, Australia where in the days of steam trains brown coal from the Latrobe Valley was evaluated for locomotive use with a negative outcome.

It is stated in [12] that there are no commercially viable bituminous coal reserves in any of the Caribbean countries, only lignite. In Jamaica peat is seen as having more potential for electricity production than lignite. The lignite in the Eastern Arctic Archipelago (next row) ranges in vitrinite reflectance from 0.19 to 0.29%. At 15 billion tonnes the quantity is immense. The comment at the beginning of the chapter that isolated places having lignite will be covered is taken to an extreme in the case of the Kerguelen Islands (11th row of the table), the 2012 population of which was 130. A point can be made in connection with this and the previous row of the table which is concerned with an Arctic location. When, as would be expected to have happened long before the 21st century, there has been a geological survey of such places it is close to being 'par for the course' that some lignite will be found. The lignite at the Kerguelen Islands is enclosed by basalt, an inorganic formation. The matter of occurrence of lignite in an inorganic formation was discussed in section 7.9. There is an Antarctic example later in the table.

The existence of $\approx$ 5 million tonnes of lignite in the Kashmir Valley was noted in section 8.2.5. That part of India more widely, the state of Jammu and Kashmir, contains tens of millions of tonnes of lignite [15,16]. Notwithstanding the Himalayan setting and very limited use, the lignite reserves of Jammu and Kashmir do feature in cartographic accounts of the lignite reserves of India.

The Lake Eyre Basin (row 14) takes in parts of Queensland, South Australia and the Northern Territory. Lignite there is adjacent to swamps (see section 7.9). To the information on Afghanistan in the following row of the table can be added the fact that there is a lignite mine at Safed Koh in the eastern part of the country [20].

In geology the term 'lens' is frequently used to mean a deposit that resembles a convex lens in shape, being thin at the edges and wide in the middle. (The word in its adjectival form was used above in the discussion of lignite from South Africa.) Lignite in Bhutan (following row) has been described as being in 'lenses' within inorganic formations comprising *inter alia* sandstone and clay [21]. The lignite close to Hudson Bay (following row) as yet has no place in the lignite production of Canada, discussed in section 6.9. Papua New Guinea (following row) is not producing coal of any rank at the present time. The lignite in the west of the country has a maceral content 96.8% in vitrinite. This deposit has been identified for possible future mining to make electricity. In Iceland (following row) there were a century ago efforts to bring the lignite reserves at Stalfjall into use [27]. There would have been both local use and export to Norway. At that time

Iceland had purchased mining rights at Spitsbergen in Norway, where the coal is bituminous in rank. Lignite in Haiti (next row) has been evaluated for briquetting [29].

There is a return to the Caribbean in the following row of the table. The point made previously that water ingress into a formation will over geological time lead to lignite formation in the form of a 'lens' leads one to expect that a mineral formation might contain lignite fragments only where water entry has been in a very small degree. This is true of the formation off Barbados referred to in the table. Such fragments at Limburg in the Netherlands (see section 5.10) have been called 'splinters' [31]. Whether in a lens or on a smaller scale as fragments, such lignite is clearly allochthonous (see section 2.6).

In Alaska (next row of the table) there is a town called Lignite, of population 0.32 million, so named because of its proximity to a lignite deposit which is called Lignite Creek. Coal from Lignite Creek is high in huminite, low-to-moderate in liptinite and low-to-moderate in inertinite [32]. Coal at Lignite Creek was discovered in 1898, just over 30 years after the USA bought Alaska from Russia. In a geological survey from about 70 years ago [33] the coal at Lignite Creek was classified as sub-bituminous C and lignite, and the total amount estimated as 0.953 billion tons (0.86 billion tonnes). Discovery of lignite whilst drilling for oil and gas at the Nicaragua coast (final row) can be related to the account in section 2.5 of the discovery of lignite beneath the Indian Ocean during hydrocarbon exploration.

23.3 Concluding remarks

Table 23.1 could be hugely increased in length, and the point made earlier in this chapter and in a previous one will be repeated with emphasis: water carrying plant debris into an inorganic formation will, over geological time, become lignite which as a result occurs in very many places. It is only where it occurs in large quantities close to a centre of population that it can be seen as an asset. By contrast small amounts of crude oil in isolated locations ('stranded oil') can be viable to recover, and there have been introduced such things as pumps and derricks which are not permanent installations but can be moved from one scene of stranded oil to another. That such measures will be taken for lignites in isolated locations is not expected.

The lignite reserves of Northern Ireland (part of the United Kingdom) were mentioned in section 5.10. Although England is certainly not an 'isolated' or 'undeveloped' place (the same is true of Alaska, discussed above), those terms do apply to its lignite deposits which are minor and have only ever been drawn on locally. They will accordingly receive brief coverage here. Lignite occurs in the Bovey basin in south Devon [35]. The Bovey basin is a source of clay for the pottery industry, and the lignite that occurs with it has been used in the kilns [36]. Beneath parts of south London there is allochthonous lignite (see section 2.6) [37]. There is also allochthonous lignite on the Isle of Wight [38]. As long ago as 1839 a discovery of lignite in the southern England county of Sussex was reported [39]. Scotland's Moray Firth is the centre of major offshore oil production. Onshore the Firth there is lignite in co-existence with sandstone and silt [40].

23.4 References

[1] http://www.factfish.com/statistic-country/south+africa/lignite+brown+coal,+additional+resources

[2] Cole D.I., Roberts D.L. 'Lignite from the western coastal plain of South Africa', *Journal of African Earth Sciences* 23: 95–117 (1996)

[3] http://www.lahmeyer.de/projekte/energie/konventionelle-stromerzeugung/single/article/yayu-coal-mine-and-thermal-power-plant-ethiopia.html

[4] http://www.guardian.co.tt/lifestyle/2014-08-24/age-coal-trinidad

[5] Pryor F.L. *Malawi and Madagascar*, Oxford University Press (1990)

[6] http://malawi.opendataforafrica.org/bglwlwd/malawi-coal-reserves

[7] http://www.nationmaster.com/country-info/profiles/Malawi/Energy

[8] Gamula G.E.T., Hui L., Peng W. 'An overview of the energy sector in Malawi', *Energy and Power Engineering* 5: 8–17 (2013)

[9] Beardsley C. *Guam Past and Present*, Tuttle Publishing (1964)

[10] Uhlig S. (ed.) *Proceedings of the XVth International Conference of Ethiopian Studies*, Otto Harrassowitz Verlag (2006)

[11] Hudson B.J. *Waterfalls of Jamaica: Sublime and Beautiful Objects*, University of the West Indies Press (2001)

[12] Goodbody I., Thomas-Hope E.M. *Natural Resource Management for Sustainable Development in the Caribbean*, Canoe Press (2002)

[13] Bustin R.M. 'Tertiary coal resources, Eastern Arctic Archipelago', *Arctic* 33: 38–49 (1980)

[14] http://www.infoplease.com/encyclopedia/world/kerguelen.html

[15] http://koausa.org/geography/chapter3.2.html

[16] Qazi S.A. *Systematic Geography of Jammu and Kashmir*, APH Publishing (2005)

[17] Davey F.J. 'The Antarctic margin and its possible hydrocarbon potential', *Tectonophysics* 114: 443–470 (1985)

[18] Habeck-Fardy A., Nanson G.C. 'Environmental character and history of the Lake Eyre Basin, one seventh of the Australian continent', *Earth Science Reviews* 132: 39–66 (2014)

[19] *The Imperial Gazetteer of India*, Trubner, London (1885)

[20] http://www.mindat.org/loc-30582.html

[21] Pareek H.S. 'Petrography and rank of the Bhangtar coals, southeastern Bhutan', *International Journal of Coal Geology* 15: 219–243 (1990)

[22] Martini I.P. 'The cold-climate peatlands of the Hudson Bay Lowland, Canada: brief overview of recent work', *Peatlands: Evolution and Records of Environmental and Climate Changes* (I.P. Martini, A. Martinez-Cortizas, W. Chesworth, eds), Elsevier (2006)

[23] Weiler P.D., Coe R.S. 'Rotations in the actively colliding Finisterre Arc Terrane: paleomagnetic constraints on Plio-Pleistocene evolution of the South Bismarck microplate, northeastern Papua New Guinea', *Tectonophysics* 316: 297–325 (2000)

[24] Weiler P.D., Coe R.S. 'Geochemistry and petrology of selected coal samples from Sumatra, Kalimantan, Sulawesi, and Papua, Indonesia', *International Journal of Coal Geology* 77: 260–268 (2009)

[25] Verhoeven K., Louwye S. 'Palaeoenvironmental reconstruction and biostratigraphy with marine palynomorphs of the Plio-Pleistocene in Tjörnes, Northern Iceland', *Palaeogeography, Palaeoclimatology, Palaeoecology* 376: 224–243 (2013)

[26] van Hoof J., van Dijken F. 'The historical turf farms of Iceland: architecture, building technology and the indoor environment', *Building and Environment* 43: 1023–1030 (2008)

[27] 'The lignite beds of Iceland', *Scottish Geographical Magazine* 33: 75 (1917)

[28] Perello-Aracena F., Hassis H.D., Rentz O. 'Future development of the energy sector in the Dominican Republic', *Energy* 15: 1029–1034 (1990)

[29] Stevenson G., Wilson T.D., Jean-Poix C., Medina N. *Coal Briquetting in Haiti: A Market and Business Assessment*, Oak Ridge National Laboratory (1987)

[30] Faugores J.C., Gonthier E., Masse L., Parra M., Pons J.C., Pujol C. 'Quaternary deposits on the South Barbados accretionary prism', *Marine Geology* 96: 247–267 (1991)

[31] Bosma H.F., Van Konijnenburg-Van Cittert J.H.A., Van der Ham R.W.J.M., Van Amerom H.W.J., Hartkopf-Fröder C. 'Conifers from the Santonian of Limburg, The Netherlands', *Cretaceous Research* 30: 483–495 (2009)

[32] Merritt R.D. 'Petrology of Tertiary and Cretaceous coals of southern Alaska', *International Journal of Coal Geology* 9: 129–156 (1987)

[33] Barnes F.F., Wahrhaftig C., Hickcox C.A., Freedman J., Hopkins D.M. *Coal Investigations in South-Central Alaska 1944–46*, US Department of the Interior (1951)

[34] Martinez Tiffer E.J., Eventov L., Chilingarian G.V. 'A review of the petroleum potential of the Caribbean margin of Nicaragua', *Journal of Petroleum Science and Engineering* 5: 337–350 (1991)

[35] Hall P.L., Angel B.R., Braven J. 'Electron spin resonance and related studies of lignite and ball clay from South Devon, England', *Chemical Geology* 13: 97–113 (1974)

[36] http://genuki.cs.ncl.ac.uk/DEV/BoveyTracey/Lignite1862.html

[37] Ellison R.A., Knox R.W.O'B., Jolley D.W., King C. 'A revision of the lithostratigraphical classification of the early Palaeogene strata of the London Basin and East Anglia', *Proceedings of the Geologists' Association* 105: 187–197 (1994)

[38] Radley J.D., Barker M.J. 'Molluscan palaeoecology and biostratinomy in a Lower Cretaceous meanderplain succession (Wessex Formation, Isle of Wight, southern England)', *Proceedings of the Geologists' Association* 111: 133–145 (2000)

[39] Mantell G.A. *The wonders of geology; or, A familiar exposition of geological phenomena; the substance of a course of lects., from notes taken by G.F. Richardson* [now a Google ebook]

[40] Mudge D.C., Copestake P. 'Revised Lower Palaeogene lithostratigraphy for the Outer Moray Firth, North Sea', *Marine and Petroleum Geology* 9: 53–69 (1992)

CHAPTER 24

NATIONAL AND INTERNATIONAL STANDARDS APPERTAINING TO BROWN COALS AND LIGNITES

24.1 Selected standards and comments

Standards relating to coal, carefully chosen, are listed in the table below. Standards are regularly reviewed and sometimes updated. When in this chapter there is reference to use of a standard in the research literature such use is not necessarily of the standard in its current form.

ASTM D388-12 sub-classifies lignites into lignite A and lignite B. The difference is on the basis of calorific value on a moist, mineral matter free basis which for Lignite B is in the range 6300 to 8300 BTU per lb (14.7 to 19.4 MJ kg^{-1}). Lignite A has a calorific value outside this range at the low end. It is noted in the rubric to the standard that it does not apply to coals high in liptinite or inertinite macerals. Such coals, it is claimed, can be identified by visual inspection from the absence of bands. The ISO standard in the following row is partly based on vitrinite reflectance, and a lignite is classified as such if it has a reflectance below 0.4%. A recently published application is to a Serbian lignite [3] with vitrinite reflectance 0.3. For an ISO standard and a national one to be one and the same as with this is not uncommon [5]. The Australian standard in the next row also uses, amongst other quantities, calorific value as a basis for classification. A 'brown coal' according to this standard has a calorific value below 19.0 MJ kg^{-1} which can be compared with the ASTM classification in the same terms.

Applications of ASTM D-3175 to lignites are legion in the research literature, and include volatile matter measurements across a range of 21 lignites from Turkey [7]. They were in the range 22.2 to 46.4% when expressed as a percentage weight of the entire coal. That having a value of 22.2% has an extremely high ash content, and the percentage volatiles on an ash-free basis is 37.4%. The ISO standard for humic acid content in the next row involves heating of a sample of the lignite for 2 hours in a sodium hydroxide solution, followed by induced precipitation of the humic material and centrifuging. It has been applied [9] to a group of 12 Greek lignites, the humic acid contents of which range from 14.3 to 40.4% of the dry coal weight. A point of possible relevance to the content of Chapter 22 is that five sub-bituminous coals, also from Greece, examined in the same study had humic acid contents in the range 9.6 to 22.2%, a range overlapping with values for the lignites.

The Indian standard in the following row is concerned with definitions. It does not distinguish between a brown coal and a lignite but gives one definition for the two. The ASTM standard in the next row applies to lignites as well as to higher rank coals. Carbon, hydrogen and nitrogen analysis are in a single instrument, approved by ASTM

STANDARD	REFERENCE
ASTM[31] D388 – 12 Standard Classification of Coals by Rank.	[1]
ISO[32] 11760:2005 Classification of coals. Also BS[33] 11760:2005.	[2]
AS[34] 2096-1987 (R2013) Classification and coding systems for Australian coals.	[4]
ASTM D-3175-11 Standard Test Method for Volatile Matter in the Analysis Sample of Coal and Coke.	[6]
ISO 5073:2013 Brown coals and lignites - Determination of humic acids.	[8]
BIS[35] IS 3810-2:2003 Solid Mineral Fuels - Vocabulary - Part 2: Terms Relating To Sampling, Testing and Analysis.	[10]
ASTM D5373 Standard Test Methods for Determination of Carbon, Hydrogen and Nitrogen in Analysis Samples of Coal and Carbon in Analysis Samples of Coal and Coke.	[11]
ASTM D4239 Standard Test Method for Sulfur in the Analysis Sample of Coal and Coke Using High-Temperature Tube Furnace Combustion.	[13]
JIS[36] M 8813:2004 Coal and coke - Determination of constituents. Also ISO 925:1997.	[15]
ASTM D2798 - 11 Standard Test Method for Microscopical Determination of the Vitrinite Reflectance of Coal.	[17]
ISO Standard 1928:2009 Solid mineral fuels - Determination of gross calorific value by the calorimeter bomb method, and calculation of net calorific value, including automated methods. *See also ČSN ISO 1928.*	[18]
DIN[37] 51719:1997 Determination of ash in solid mineral fuels.	[20]
ASTM D6542 - 05(2010) Standard Practice for Tonnage Calculation of Coal in a Stockpile.	[22]
PSQCA[38] 935-1989 Brown Coal and Lignites, Determination of the Yield of Toluene Soluble Extract.	[23]

Standard	Reference
ISO 5069:1983 Brown coals and lignites - Principles of sampling Part 1: Sampling for determination of moisture content and for general analysis. Part 2: Sample preparation for determination of moisture content and for general analysis.	[24]
CEN[39]/TR 15840:2009 Evaluation of conformity of fly ash for concrete.	[26]
Standard: GOST[40] 25543 Brown coals, hard coals and anthracites: Classification according to technological parameters.	[27]
ASME[41] PTC 34 (2007) Waste Combustors with Energy Recovery.	See section 6.6, and qualifying footnote.
PN[42] EN 14658:2009 Continuous Handling Equipment And Systems - General Safety Requirements For Continuous Handling Equipment For Opencast Lignite Mining.	[30]
ČSN[43] ISO 1928 Solid mineral fuels — Determination of gross calorific value by the bomb calorimetric method and calculation of net calorific value.	[31]
GB[44]/T 2559-2005 Analysis of lignite wax.	[34]
ISO 602:2015 Coal - Determination of mineral matter.	[35]
SABS[45]/T C 027/80 06 Coal classification.	[36]
BS[46] 1016-112:1995 Methods for analysis and testing of coal and coke. Determination of Hardgrove grindability index of hard coal.	[37]
BS EN 14034-1:2004 Determination of explosion characteristics of dust clouds. Determination of the maximum explosion pressure P_{max} of dust clouds.	[38]

(e.g., [12]), which works by burning the coal at 950°C. Water (from the coal hydrogen) and carbon dioxide (from the coal carbon) are measured by infrared, and nitrogen by a thermal conductivity detector. The instrument is subjected to calibrations checks with model compounds such as ethylene diamine tetra acetic acid (EDTA) at intervals specified in the standard [12]. Full analysis by element ('ultimate analysis') also requires that sulphur be determined, and this is in any case important because of sulphur dioxide emissions. This is by means of ASTM D4239 (following row), which can also be applied across a range of rank. It involves heating of the coal sample in a tube furnace at 1350°C under flowing oxygen, to convert all of the coal sulphur to oxides of sulphur which can be quantitatively determined. Both of these standards were applied in published work on a North Dakota lignite [14], for which the results were as follows: C 79.3%, H 3.8%, N 1.2%, S 1.6%, O by difference 14.1%, all dry ash-free basis. Oxygen is always by difference in such analyses.

The Japanese standard is applied to two 'standard coals' from the Japan Coal Energy Center [16] to give a proximate analysis for each. It is by means of another Japanese standard – JIS M1002 – that the two standard coals are classified as being brown coals. The ASTM standard for vitrinite reflectance in the following row is also for higher rank coals as well as for lower rank ones. A recent published application [18] is to a Texan lignite, reflectance values for multiple samples of which were in the range 0.22 to 0.47%.

The ISO standard in the following is concerned with calorific values, and in its ISO guise has in recently published work [19] been applied to an Indonesian lignite the calorific value of which was 19.8 MJ kg^{-1} when the coal moisture was 27%. The German standard in the next row has been applied to braunkohle from two mines in Bavaria [21]. Coal from one had ash in the range 13.2 to 48.6% depending on the position in the bed from which the sample was taken. The other had ash contents from 4.3 to 34.6%.

The ASTM standard in the next row is for monitoring the amount of coal remaining in a stockpile after removals for sale. It applies to all ranks of coal and is based on stockpile geometry. The moisture content of the coal in a particular stockpile is factored in. The PSQCA standard in the following row is one of a number issued by that body relating to coal science and utilisation. Another is PSQCA 757-1989 'Method for Determination of Yield Tar Water, Gas and Coke Residue by Low-Temperature Distillation of Brown Coal and Lignite'.

The ISO standard in the following row, in two parts, is invoked in very recent work on lignite from the Most basin in the Czech Republic (see section 5.1.6) [25]. GOST is the standards body of countries in the Former Soviet Union where there is huge lignite production as described in Chapter 9. GOST 25543 has recently been applied to lignite from the Moscow region [28]. Like its ASTM counterpart (row 1), GOST 25543 provides for classifications within coal rank [29] and the Russian lignite in [28] is assigned descriptor 2B.

The Polish standard in the following row is also a European standard and is sometimes cited either as that alone or with the affiliation of another European country, for example as DIN EN 14658:2009 or as UNE[47] EN 14658:2009. The standard is concerned with the conveyance of lignite and also of overburden when removal of that is taking place. Movement of equipment such as that shown on the book cover comes within the scope of the standard.

The Czech standard in the next row is also an ISO standard as indicated. Applications are many, and include on to a New Zealand lignite [32] the lower heat value (disregarding the heat released by condensation of water in the combustion products) of which was measured as 17.3 MJ kg^{-1}. Within the Czech Republic the standard has been applied, with its ČSN prefix, to biomass [33].

Over the remaining rows of the table the South African and British standards bodies are introduced. In section 1.3.2 it is stated that the Hardgrove index has become the basis of national standards. The entry in the final row is an example of this.

24.2 References

[1] http://www.astm.org/Standards/D388.htm

[2] https://www.iso.org/obp/ui/#iso:std:iso:11760:ed-1:v1:en

[3] Životić D., Stojanović K., Gržetić I., Jovančićević B., Cvetković O., Šajnović A., Simić V., Stojaković R., Scheeder G. 'Petrological and geochemical composition of lignite from the D field, Kolubara basin (Serbia)' International Journal of Coal Geology 111 5-22 (2013).

[4] http://infostore.saiglobal.com/store/details.aspx?ProductID=258517

[5] Jones J.C. 'The standards bodies and applied chemistry' Chemistry in Australia March 2013 p. 28.

[6] http://infostore.saiglobal.com/store/Details.aspx?ProductID=1464450&gclid=CPis--y

F6MUCFRAJvAod-6kAUA

[7] Kucukbayrak S. 'Influence of the mineral matter content on the combustion characteristics of Turkish lignites' Thermochimica Acta 216 119-129 (1993).

[8] http://www.iso.org/iso/iso_catalogue/catalogue_ics/catalogue_detail_ics.htm

?csnumber=20458

[9] Giannouli A., Kalaitzidis S., Siavalas G., Chatziapostolou A., Christanis K., Papazisimou S., Papanicolaou C., Foscolos A. 'Evaluation of Greek low-rank coals as potential raw material for the production of soil amendments and organic fertilizers' International Journal of Coal Geology 77 383-393 (2009).

[10] http://infostore.saiglobal.com/store/details.aspx?ProductID=1050562

[11] http://www.astm.org/Standards/D5373.htm

[12] 'Carbon, Hydrogen, and Nitrogen in Coal' Application note, LECO Corporation, Saint Joseph MI.

[13] http://www.astm.org/Standards/D4239.htm

[14] Mangena S.J., Bunt J.R., Waanders F.B. 'Physical property behaviour of North Dakota lignite in an oxygen/steam blown moving bed gasifier' Fuel Processing Technology 106 326-331 (2013).

[15] http://infostore.saiglobal.com/store/details.aspx?ProductID=773391

[16] Kashiwakura S., Takahashi T., Nagasaka T. 'Vaporization behavior of boron from standard coals in the early stage of combustion' Fuel 90 1408-1415 (2011).

[17] http://www.astm.org/Standards/D2798.htm

[18] http://www.iso.org/iso/iso_catalogue/catalogue_tc/catalogue_detail.htm?csnumber=41592

[19] Cheng J., Zhou F., Wang X., Liu J., Zhou J., Cen K. 'Physicochemical properties of Indonesian lignite continuously modified in a tunnel-type microwave oven for slurribility improvement' Fuel 150 493-500 (2015).

[20] http://webstore.ansi.org/RecordDetail.aspx?sku=DIN+51719%3A1997

[21] Dehmer J. 'Petrographical and organic geochemical investigation of the Oberpfalz brown coal deposit, West Germany' International Journal of Coal Geology 11 273-290 (1989).

[22] http://www.astm.org/Standards/D6542.htm

[23] http://www.psqca.com.pk/e-catelog/disprec.asp?disprec=935-1989

[24] http://www.iso.org/iso/home/store/catalogue_ics/catalogue_detail_ics

.htm?ics1=73&ics2=40&ics3=&csnumber=11091

[25] Havelcová M., Sýkorová I., Mach K., Trejtnarová H., Blažek J. 'Petrology and organic geochemistry of the lower Miocene lacustrine sediments (Most Basin, Eger Graben, Czech Republic)' International Journal of Coal Geology 139 26-39 (2015).

[26] http://standards.cen.eu/dyn/www/f?p=204:105:0

[27] http://runorm.com/product/view/2/11087

[28] Kus J. 'Application of confocal laser-scanning microscopy (CLSM) to autofluorescent organic andmineral matter in peat, coals and siliciclastic sedimentary rocks — A qualitative approach' International Journal of Coal Geology 137 1-18 (2015).

[29] http://pubs.usgs.gov/of/2001/ofr-01-104/fsucoal/html/readme.htm

[30] http://infostore.saiglobal.com/EMEA/Portal.aspx?publisher=PKN

[31] http://csnonlinefirmy.unmz.cz/html_nahledy/44/42115/42115_nahled.htm

[32] Saw W.L., Pang S. 'Co-gasification of blended lignite and wood pellets in a 100 kW dual fluidised bed steam gasifier: The influence of lignite ratio on producer gas composition and tar content' Fuel 112 117-124 (2013).

[33] Brant V, Pivec J., Fuksa P., Necka′ K., Kocourkova′ D., Venclova V. 'Biomass and energy production of catch crops in areas with deficiency of precipitation during summer period in central Bohemia' Biomass and Bioenergy 35 1286-1294 (2011)

[35] http://shop.bsigroup.com/ProductDetail/?pid=000000000030314896

[36] https://www.sabs.co.za/

[37] http://shop.bsigroup.com/ProductDetail/?pid=000000000000525581

[38] http://shop.bsigroup.com/ProductDetail/?pid=000000000030237579

CHAPTER 24 ENDNOTES

31 American Society for Testing and Materials. HQ in Philadelphia PA.

32 ISO. HQ in Geneva.

33 British Standards Institution. HQ in London.

34 Standards Australia. HQ in Sydney.

35 Bureau of Indian Standards. HQ in New Delhi.

36 Japanese Industrial Standards. HQ in Tokyo.

37 Deutsches Institut für Normung. HQ in Berlin.

38 Pakistan Standards and Quality Control Authority. HQ in Karachi.

39 Comité Européen de Normalisation, HQ in Brussels.

40 Gosudarstvennye Standart State. HQ in Moscow.

41 Formerly American Society of Mechanical Engineers now called just ASME. HQ in New York.

42 Polish Committee for Standardisation. HQ in Warsaw.

43 Czech Standards Institute. HQ in Prague.

44 Guobiao standard, issued by Standardization Administration of China.

45 South African Bureau of Standards. HQ in Pretoria.

46 British Standards Institution. HQ in London.

47 Una Norma Española. HQ in Madrid.

INDEX

A

B

C

R

S

T

U

V

W

Y